Mark Anthony Benvenuto

Materials Chemistry

Also of Interest

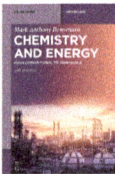

Chemistry and Energy.
From Conventional to Renewable
Benvenuto, 2025
ISBN 978-3-11-133090-7, e-ISBN 978-3-11-133109-6

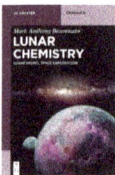

Lunar Chemistry.
Lunar Mining, Space Exploration
Benvenuto, 2025
ISBN 978-3-11-138712-3, e-ISBN 978-3-11-138730-7

Engineering Materials Characterization
Kumar, Zindani, 2024
ISBN 978-3-11-099760-6, e-ISBN 978-3-11-099759-0

Manufacturing Processes.
Metal Forming, Machining, Metal Casting, Welding and Additive Manufacturing
Alam, Dwivedi, Srivastava, Mishra, 2025
ISBN 978-3-11-914753-8, e-ISBN 978-3-11-220589-1

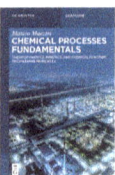

Chemical Processes Fundamentals.
Thermodynamics, Kinetics, and Chemical Reaction Engineering Principles
Maestri, 2026
ISBN 978-3-11-172286-3, e-ISBN 978-3-11-172318-1

Mark Anthony Benvenuto

Materials Chemistry

Chemical Bonding, Structure of the Matter, Materials, Types of Materials

2nd Edition

DE GRUYTER

Author
Prof. Dr. Mark Anthony Benvenuto
Department of Chemistry & Biochemistry
University of Detroit Mercy
4001 W. McNichols Rd.
DETROIT MI 48221-3038
United States of America
benvenma@udmercy.edu

ISBN 978-3-11-914739-2
e-ISBN (PDF) 978-3-11-220582-2
e-ISBN (EPUB) 978-3-11-220616-4

Library of Congress Control Number: 2025942349

Bibliographic information published by the Deutsche Nationalbibliothek
The Deutsche Nationalbibliothek lists this publication in the Deutsche Nationalbibliografie;
detailed bibliographic data are available on the Internet at http://dnb.dnb.de.

© 2026 Walter de Gruyter GmbH, Berlin/Boston, Genthiner Straße 13, 10785 Berlin
Cover image: Irina Vodneva/iStock/Getty Images Plus
Typesetting: Integra Software Services Pvt. Ltd.

www.degruyterbrill.com
Questions about General Product Safety Regulation:
productsafety@degruyterbrill.com

Preface

The broad field of materials chemistry or materials science is a constantly evolving and growing one, and thus one that defies having what might be called fixed, pre-determined, or defined edges. In the past it has been considered an area of applied solid state chemistry; but today there are some liquids and even gases that can be considered part of materials chemistry. This book contains sections and chapters on materials that are considered old, established and defined – subjects that will be found in virtually all materials chemistry books – but also on subjects that are new and rapidly expanding.

No book on materials chemistry would be complete without some discussion of metals, plastics, ceramics, glasses and cement. These, plus a few other traditional materials, are still widely used today. For example, numerous species of wood are widely used in the modern world, even as wood composites, wood laminates, and artificial wood or plastic wood find increasing applications and uses in our economy. Yet some materials that have been known for decades have found increasing uses and new applications over time. The perfect example might be neodymium, first discovered and reported by Carl Auer von Weisbach in 1885, first isolated and purified in 1925, considered a curiosity for decades, but now used as a component of supermagnets. Thus, the element is a component in the billions of cellular phones in service throughout the world.

Of course, there are some thoroughly modern materials that should be discussed in any materials chemistry book. Subjects such as semi-conductors come quickly to mind, even though experts in the field might no longer consider them to be entirely modern – with the earliest patent for one now being more than 100 years old.

More modern materials would certainly include superconductors, a field that is still growing both in terms of materials which exhibit the phenomenon, and in terms of the theoretical background behind superconductivity at various temperatures. The superconducting mag-lev trains that thirty years ago were a promise for the future remain just that, a promise for some undetermined future. But this application as well as others may now be much closer.

While different books on materials science and materials chemistry will routinely cover several topics that are now considered common, and that are considered widespread in terms of established applications, many do not yet mention metal-organic frameworks – aka MOFs – or metal foams, or bio-based materials such as mycelium. I have tried to incorporate information and discussion on these topics, as it is probably fair to say they represent some of the newest classes of materials that are finding a broad possible range of uses.

Thank you.

https://doi.org/10.1515/9783112205822-202

Additionally, when writing a book like this, I need to thank several people, all of whom have helped answer my questions as this work was put together. Klaus Friedrich, Matt Mio, Kate Lanigan, Kendra Evans, Mary Lou Capsrs, Jon Stevens, Sue White, and Heinz Plaumann, thank you all for giving me tips and advice, even when you did not know it was for this book, and thought it was just another one in my never-ending string of random questions. I also have to thank Jane Schley and Meghann Murray for their help at many points along the way. My editors, Ria Sengbusch and Karin Sora, have both been tremendously helpful and patient as this project has come to fruition. Thank you both. Finally, I must thank my wife Marye, and my sons David and Christian for putting up with all the griping and moaning I do when I'm in the midst of a project, a trait I suspect I share with many authors.

Contents

Chapter 4
Chemical bonding —— 37

Chapter 5
Structure of matter —— 52

Chapter 6
Molecules —— 63

Chapter 1
Overview and introduction

1.1 Introduction and history: the roots of materials science

"Materials science" and "materials chemistry" are two terms for an interdisciplinary aspect of science or nexus of traditionaly different sciences that deal with the behavior of a wide variety of substances, often in the solid state. There are materials that can be called traditional, which means they have been created, produced on a large scale, often harvested, and used in one way or another, for centuries or even millenia. But there are also new materials that have been used for much shorter times. Examples of traditional materials are wood, stone, various metals, clay, glass, and bone. New materials would certainly include almost all plastics, vulcanized rubber, bone mimics, an enormous variety of metal alloys, and metal-organic frameworks, to name just a few of many. A materials chemistry book seeks to examine such materials, makes comparisons among them, discusses their current uses, and perhaps suggests future uses for such materials.

Table 1.1 shows a nonexhaustive list of different classes of materials and gives examples of their uses, as well as an association or society devoted to them.

Table 1.1: Traditional and modern materials.

Material	Traditional or new	Example, uses	Representative association
Wood	Traditonal	Construction, ships	North American Wholesale Lumber Association [1]
Metals	Traditonal	Tools, weapons	American Iron and Steel Institute [7]
Stone	Traditonal	Building	Natural Stone Institute [12]
Clay	Traditonal and new	Pottery, record keeping, insulators	The Clay Minerals Society [17]
Glass	Traditional	Tableware, windows	National Glass Association [22]
Leather	Traditional	Clothing, armor, saddles	U.S. Hide, Skin, and Leather Association (USHSLA) [27]
Plastics	New	Piping, containers (an enormous number of uses)	Plastics Industry Association [32]
Rubber	New	Tires	Association for Rubber Products Manufacturers [37]

https://doi.org/10.1515/9783112205822-001

Table 1.1 (continued)

Material	Traditional or new	Example, uses	Representative association
Metal-organic frameworks	New	Catalyst, gas storage	Metal-Organic Frameworks – International Commission [42]
Catalysts	New	Pollution abatement, wide variety of uses	North American Catalysis Society [43]
Semiconductors	New	Controlled electrical needs	Semiconductor Industry Association [46]
Biomaterials	New	Hard and soft tissue, others	Society for Biomaterials [51]
Energy materials	New	Photovolatics, fuel cells, batteries	Basic Energy Sciences [55]
Nanomaterials	New	Fullerenes	National Nanotechnology Initiative [58]

It should be noted that the organizations mentioned in the right-most column of Table 1.1 are only examples of some society or association that can claim to represent a field. There are numerous trade organizations for most of the categories: some national and some multinational [1–58].

1.2 New materials as the periodic table was formed

For much of the history of humanity, a relatively few materials served the needs of people everywhere. To illustrate a broad example, choose cloth and clothing: cotton, wool, linen, and silk tended to be the types of cloth all people wore, but that were made from only a few sources. Cotton is from the *Gossypium* plant; wool comes from sheep, goats, llamas, alpacas, and a few other species of animals; linen is made from flax (*Linum usitatissimum*); and silk is from the *Bombyx mori* silkworm in its larval state. Leather is also a traditional clothing material, and is also used for shoes and armor, before it became the luxury item it tends to be today. Leather is the chemically treated skins of various animals.

A person can make a number of arguments for how time and human activity can be divided. But one that certainly works as far as materials are concerned is connected to the veritable explosion of new elements that were discovered in the late 1700s and early 1800s. Figure 1.1 shows the periodic table of elements with only the generally accepted year of the discovery of each. Several listed as "Anc." were known in ancient times. We have not included any elements past rutherfordium because all such have been produced and isolated in extremely tiny amounts.

1776																	1895
1817	1797											1808	Anc.	1772	1774	1886	1898
1807	1755											1825	1824	1669	Anc.	1774	1894
1808	1879	1791	1830	1797	1774	1774	Anc.	1751	1751	Anc.	1875	1886	1886	1817	1826	1825	1898
1861	1790	1794	1789	1801	1781	1937	1844	1803	1803	Anc.	1817	1863	Anc.	Anc.	1782	1811	1898
1860	1808	1839	1923	1802	1783	1925	1803	1803	1735	Anc.	Anc.	1861	Anc.	1400	1898	1940	1900
1939	1898	1899	1964														

1803	1885	1885	1945	1879	1901	1880	1843	1886	1867	1842	1879	1878	1907
1829	1913	1789	1940	1940	1944	1944	1949	1950	1952	1952	1955	1958	1961

Figure 1.1: Periodic table of the elements and their date of discovery.

In Figure 1.1, notice how many elements were discovered in the nineteenth century. This gave rise to the possibility of many new alloys, which in turn could be put to a wide variety of uses. Again, choosing one example, an alloy called German silver was produced by Dr. Lewis Feuchtwanger – a German expatriate to the United States – as early as 1834. The alloy is one of copper–nickel–zinc–tin, with some differences in what Dr. Feuchtwanger used over time. He tried to convince the United States Congress to authorize this alloy for base metal coins of the United States, such as 1¢ or 2¢ pieces. Although his idea was turned down, the United States 5¢ piece ultimately shifted in the year 1866 from being a silver–copper alloy coin to one made of 75% copper and 25% nickel – an alloy still used today. Despite his setback, Dr. Feuchtwanger produced and distributed his own tokens made of his preferred alloy, still called Feuchtwanger cents today. An example of one of these tokens is shown in Figure 1.2.

Another example of a relatively modern alloy, one that finds its use today, is the production of Wood's metal fusible alloy, which is a mixture of 50% bismuth, 26.7% lead, 13.3% tin, and 10% cadmium (as weight percents). First made by Dr. Barnabas Wood, who was a dentist as well as an inventor, the alloy melts at 70 °C and finds its use today as a solder. Figure 1.3 shows ingots of Wood's metal, a common form in which it can be purchased.

A third example of an alloy that is relatively new, and yet that finds wide use today is the neodymium–iron–boron type of magnets. The alloy, sometimes written as $Nd_2Fe_{14}B$, was developed by two companies independently: Sumitomo Special Metals and General Motors. These magnets have found wide use and applications today. Perhaps the best known is their presence in cellular phones – in several different spots within cell phones. Figure 1.4 shows a common modern cell phone, an item far

Figure 1.2: Example of a Feuchtwanger cent.

Figure 1.3: Wood's metal ingots.

smaller today than any phone made even as recently as the 1980s, in part because of the presence of such alloy magnets.

Also, note that in Figure 1.1, the actinides and the noble gases are the only groups we might claim have been discovered systematically, although some scientists will claim that for the actinides it was not a matter of discovery, but rather one of progressive, directed syntheses.

Figure 1.4: Modern cell phone size.

1.3 The materials chemistry explosion, post-Second World War

The Second World War is far enough behind us in time that the majority of people alive today have no memory of it. And yet the Second World War was a time of crisis that produced a rather amazing variety of innovations, including new materials, often as some substitute use for an older, established material. Consider the following two perhaps obvious examples. First, synthetic rubber – used widely during the war because the Hevea brasiliensis trees which produced natural rubber latex were mostly in territories controlled by Imperial Japan. Second, parachute silk – originally made from natural silk, but later made from the polyamide fiber nylon, manufactured at DuPont – again, because the lands where silkworms were cultivated were either in Japanese hands, or hard to get to. Synthetic rubber now finds a host of uses, and derivatives of it have been discovered and produced for decades. Figure 1.5 shows the Lewis structiure of the repeat isoprene unit of synthetic rubber once polymerized, as well as that of synthetic rubber, the latter made from 1,3-butadiene. Likewise, nylon – the repeat unit of which is shown in Figure 1.6 – has become a massive bulk chemical,

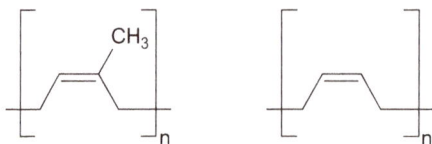

Figure 1.5: Lewis structure, natural and synthetic rubber.

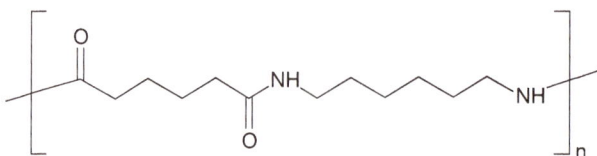

Figure 1.6: Lewis structure, repeat unit of nylon.

and it is probably fair to say that hundreds of derivatives of it have been produced and their properties examined.

1.4 Recycling

The idea of producing an item from any material, including those which are fairly new, and recycling it for some second use, is both an idea that gets connected with materials chemistry, and is another idea we might claim arose during the Second World War – although, in the United States, the First World War was the first time people saw gathering and re-use of materials such as metals on a large, national scale. While there was a period in the late 1940's and early 1950's of what can be called peace and plenty, in which objects were made, used, and thrown away, it eventually became apparent that many of these materials and the user end items made from such materials would last for centuries or even millenia if they were simply discarded into landfills – they were afterall designed to last, and not to degrade or rot. The presence of so much obvious, lasting pollution led to calls for some kind of wiser use of materials and parallel treatment of our environment. Clearly, some actions had to be taken to conserve resources, and to use materials wisely.

The struggle to use and re-use materials wisely continues today, where in general people recognize that recycling is a smarter use of any resource or any material than a single use followed by permanent disposal. Often, the incentive is not environmental, but rather, economic. Aluminum beverage cans, such as those in Figure 1.7, are an excellent example of the costs saved when a metal is recycled. Some sources claim that recycling an aluminum can is actually 99% more energy efficient than refining virgin aluminum from bauxite [10]. Many public places now have recycling receptables specifically for aluminum, as well as other materials, such as that shown in Figure 1.8.

Figure 1.7: Common, recyclable aluminum beverage cans.

Figure 1.8: Public recycling container for aluminum and plastic.

But aluminum is not the only metal, nor indeed the only material, that is recycled on a large scale. Steel, iron, copper, brass, bronze, precious metals, glass, paper, plastics, and rubber are also recycled widely. As well, kiosks have been placed in public spots such as malls for the recycling of old cell phones.

Sometimes recycling is done for some economic reason – such as a bottle return refund – and sometimes it is mandated by a local, regional, or national government. For example, Figure 1.9 shows the common recycling triangle, and what is called the RIC – the resin identification code – on the bottom of a polypropylene container. When sorted by RIC, several plastics are relatively easy to recycle, and such recycling can be economically feasible.

While it sounds impressive that some countries have advanced to a set of national recycling programs, the size of the nation in question becomes a factor in the feasibility of such programs on a global scale. A fairly obvious example might be that it is much easier to enact a national metal recycling program in Denmark, a nation of approximately 5.6 million people, as opposed to the United States, which has roughly 327 million people, or the People's Republic of China, with almost 1.4 billion people.

Despite size differences though, many states, provinces, and countries have taken some action to ensure the recycling and re-use of several different classes of materials. Recycling has become so pervasive that there are numerous organizations devoted to it [59–68].

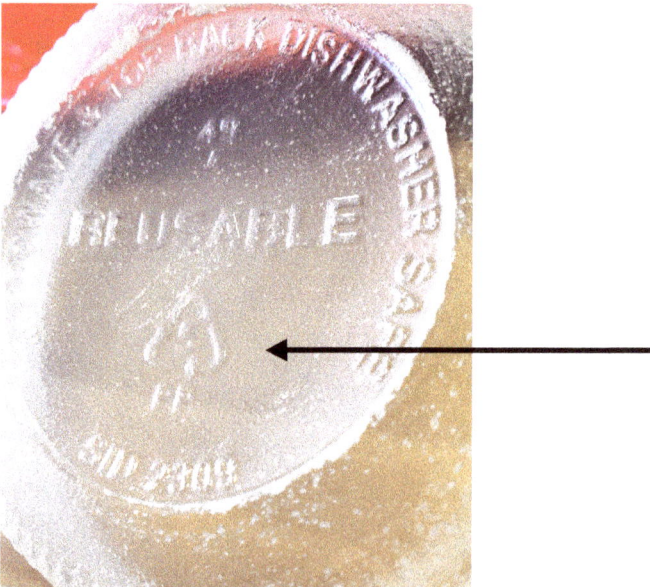

Figure 1.9: Example of an RIC, used for recycling purposes.

References

[1] North American Wholesale Lumber Association. Website. (Accessed 8 June 2025, as: https://nawla.org).

[2] National Hardwood Lumber Association. Website. (Accessed 8 June 2025, as: https://nhla.com).

[3] Forest Products Association of Canada. Website. (Accessed 8 June 2025, as: https://www.fpac.ca).

[4] Canadian Lumber Standards Accreditation Board. Website. (Accessed 8 June 2025, as: https://www.clsab.ca).

[5] APAwood-Europe. Website. (Accessed 8 June 2025, as: https://apawood-europe.org).

[6] Australian Forest Products Association. Website. (Accessed 8 June 2025, as: https://ausfpa.com.au).

[7] American Iron and Steel Institute. Website. (Accessed 21 March 2020, as: https://www.steel.org).

[8] Nickel Institute. Website. (Accessed 8 June 2025, as: https://nickelinstitute.org).

[9] Copper Development Association. Website. (Accessed 8 June 2025, as: https://www.copper.org).

[10] The Aluminum Association. Website. (Accessed 8 June 2025, as: https://www.aluminum.org).

[11] European Steel Association. Website. (Accessed 8 June 2025, as: https://www.eurofer.eu).

[12] Natural Stone Institute. Website. (Accessed 21 March 2020, as: https://www.naturalstoneinstitute.org).

[13] Ontario Stone, Sand & Gravel Association. Website. (Accessed 8 June 2025, as: https://www.ossga.com).

[14] Stone Federation GB. Website. (Accessed 8 June 2025, as: https://www.stonefed.org.uk).

[15] Euroroc. Website. (Accessed 8 June 2025, as: https://www.euroroc.net).

[16] Australian Stone Advisory Association. Website. (Accessed 8 June 2025, as: https://asaa.com.au).

[17] The Clay Mineral Society. Website. (Accessed 21 March 2020, as: https://www.clays.org).

[18] The American Ceramic Society. Website. (Accessed 8 June 2025, as: https://ceramics.org).

[19] The Ontario Clay and Glass Association. Website. (Accessed 8 June 2025, as: https://www.clayandglass.on.ca).

[20] The European Ceramic Industry Association. Website. (Accessed 8 June 2025, as: https://cerameunie.eu).

[21] The Australian Ceramics Association. Website. (Accessed 8 June 2025, as: https://australianceramics.com).

[22] National Glass Association. Website. (Accessed 21 March 2020, as: https://www.glass.org).

[23] The Stained Glass Association of America. Website. (Accessed 8 June 2025, as: https://stainedglass.org).

[24] Glass Association of North America. Website. (Accessed 8 June 2025, as: https://www.glassglobal.com).

[25] Glass Alliance Europe. Website. (Accessed 8 June 2025, as: https://glassallianceeurope.eu).

[26] Australian Glass and Window Association. Website. (Accessed 8 June 2025, as: https://www.agwa.com.au).

[27] United States Hide, Skin, and Leather Association (USHSLA). Website. (Accessed 8 June, 2025 as: https://www.ushsla.org).

[28] Leather and Hide Council of America. Website. (Accessed 8 June 2025, as: https://www.usleather.org).

[29] Luggage, Leathergoods, Handbags & Accessories Association of Canada. Website. (Accessed 8 June 2025, as: https:www.worldfootwear.com/organizations/llha628html).

[30] Leather UK. Website. (Accessed 8 June 2025, as: https://leatheruk.org).

[31] Euroleather. Website. (Accessed 8 June 2025, as: https://www.euroleather.net).

[32] Plastics Industry Association. Website. (Accessed 8 June 2025, as: https://www.plasticsindustry.org).

[33] The Association of Plastics Recyclers. Website. (Accessed 8 June 2025, as: https://plasticsrecy cling.org).

[34] Canadian Plastics. Website. (Accessed 8 June 2025, as: https://www.canplastics.com).

[35] British Plastics Federation. Website. (Accessed 8 June 2025, as: https://www.bpf.co.uk).

[36] Plastics Europe. Website. (Accessed 8 June 2025, as: https://plasticseurope.org).

[37] Association for Rubber Products Manufacturers. Website. (Accessed 8 June 2025, as: https://arp minc.com).

[38] U.S. Tire Manufacturers Association. Website. (Accessed 8 June 2025, as: https://www.ustires.org).

[39] Tire and Rubber Association of Canada. Website. (Accessed 8 June 2025, as: https://tracanada.ca).

[40] British Rubber Manufacturers' Association. Website. (Accessed 8 June 2025, as: https://www.bpf. co.uk).

[41] European Tyre & Rubber Manufacturers Association, ETRMA. Website. (Accessed 8 June 2025, as: https://www.etrma.org).

[42] Metal-Organic Frameworks – International Commission. Website. (Accessed 8 June 2025, as: https:// mof-international.org).

[43] North American Catalysis Society. Website. (Accessed 8 June 2025, as: https://nacatsoc.org).

[44] IACS – The International Association of Catalysis Societies. Website. (Accessed 8 June 2025, as: http://iacs-catalysis.org).

[45] EFCATS – European Federation of Catalysis Societies. Website. (Accessed 8 June 2025, as: https://ef cats.org).

[46] Semiconductor Industry Association. Website. (Accessed 8 June 2025, as: https://www.semiconduc tors.org).

[47] SEMI. Website. (Accessed 8 June 2025, as: https://www.semi.org).

[48] European Semiconductor Industry Association (ESIA). Website. (Accessed 8 June 2025, as: https:// www.eusemiconductors.eu).

[49] SEAJ. Semiconductor Equipment Association of Japan. Website. (Accessed 8 June 2025, as: https:// www.seaj.or.jp).

[50] KSIA. Korea Semiconductor Industry Association. Website. (Accessed 8 June 2025, as: http://www. sedex.org).

[51] Society for Biomaterials. Website. (Accessed 8 June 2025, as: https://biomaterials.org).

[52] ESB. European Society for Biomaterials. Website. (Accessed 8 June 2025, as: https://www.esbiomate rials.eu).

[53] Canadian Biomaterials Society. Website. (Accessed 8 June 2025, as: https://biomaterials.ca).

[54] Australasian Society for Biomaterials and Tissue Engineering. Website. (Accessed 8 June 2025, as: https://www.asbte.org).

[55] Basic Energy Sciences. Website. (Accessed 8 June 2025, as: https://www.energy.gov/science/bes/ basic-energy-sciences).

[56] Energy Sciences Coalition. Website. (Accessed 8 June 2025, as: https://www.aplu.org).

[57] European Energy Research Alliance. Website. (Accessed 8 June 2025, as: https://www.eera-set.eu).

[58] National nanotechnology Initiative. Website. (Accessed 8 June 2025, as: https://www.nano.gov).

[59] The Recycling Partnership. Website. (Accessed 8 June 2025, as: https://recyclingpartnership.org).

[60] Keep America Beautiful. Website. (Accessed 8 June 2025, as: https://kab.org).

[61] National Recycling Coalition. Website. (Accessed 8 June 2025, as: https://nrcrecycles.org).

[62] Canadian Association of Recycling Industries. Website. (Accessed 8 June 2025, as: https://cari-acir.org).

[63] The Recycling Association. Website. (Accessed 8 June 2025, as: http://www.therecyclingassocia tion.com).

[64] EuRIC – Advocating Recycling in Europe. Website. (Accessed 8 June 2025, as: https://circularecon omy.europa.eu).

[65] Australian Council of Recycling. Website. (Accessed 8 June 2025, as: https://www.acor.org.au).

[66] Recycling Association of South Africa. Website. (Accessed 8 June 2025, as: https://raa.africa).

[67] The Japan Containers and Packaging Recycling Association. Website. (Accessed 8 June 2025, as: https://www.jcpra.or.jp/english).

[68] Kora – Korea Resource Circulation Service Agency. Website. (Accessed 8 June 2025, as: http://www. kora.or.kr).

Chapter 2
Basic principles of materials chemistry

2.1 Solid-state materials

The basic ideas of materials chemistry often involve solids and solid-state materials. The term *solid-state materials* encompasses everything from elemental solids – many of them were metals – to alloys, to crystalline ionic solids, to molecular solids, as well as to other solid materials. Oftentimes, these solids are produced and created by design. The term also has become associated in recent decades with what are called *nonmolecular solids*. Whatever the case, though, materials in their solid state are a central idea of materials chemistry.

These solid-state materials fall into several categories, most of which will have part of a chapter or an entire chapter devoted to them later in this book. They include:
1. Crystals, which are the subject of Chapter 5.
2. Plastics, which are the subject of Chapter 7.
3. Alloys, which are treated in Chapter 8.
4. Magnetic materials, which are also treated in Chapter 8.
5. Semiconductors, which are the subject of Chapter 11.
6. Superconductors, which are the subject of Chapter 12.
7. Glasses and ceramics, which are the subject of Chapters 14 and 15.

Table 2.1 presents a nonexhaustive list of these broad categories of solid-state materials, as well as one or two examples of their traditional use, plus some examples of a modern use [1]. The categories may seem somewhat random, as some of them are defined by the composition of a material, while others are defined by a characteristic. An example of the first would be alloys – mixtures of metals, and at times nonmetals. The starting elements, and the overall combination of elements, are what define an alloy. And an example of the second might be the category of superconductors. In this case, it does not matter what the material is, or whether it is a single element or multiple, or whether it is metallic or nonmetallic. Only the characteristic of superconductivity – the complete loss of resistivity at some temperature – defines a material as being a superconductor.

2.1.1 Metals and alloys

It is fair to say that the roots of modern metal and alloy chemistry lies in a few examples that reach into the ancient past. These metals were pursued and purified for the properties they possessed – usually hardness, ductility, and malleability. All those used in ancient times are still in some way used today [2].

https://doi.org/10.1515/9783112205822-002

Table 2.1: Solid-state materials.

Name	Traditional example and application	Modern example and application	Comments
Alloys	Bronze or steel; tools and weapons	Niobium steel, jet engines Aluminum, airplane components	
Ceramics	Pottery, containers used since ancient times	Bioceramics, dental implants	
Crystals	Gems, indicators of wealth and status since ancient times	Synthetic diamonds, as jewelry and industrial abrasives.	
Magnets	Lodestone, historically used as compasses	NdFeB magnet, used in electric motors	Modern magnets are much stronger than traditional ones
Semiconductors	"Cat whiskers," detecting radio waves	Silicon, in electronic devices	
Superconductors	Discovered in 1911, mercury	Magnets, used in NMR and MRI	Continued research is aimed at finding room-temperature superconductors

Bronze is a copper–tin alloy that has been known for millennia, but that is now made in so many different varieties, and with so many different elements beyond just copper and tin, that the American Society for Testing and Materials (ASTM) has an entire set of codes for different bronzes.

Another metal that has its roots in antiquity is steel. In the distant past, steel was made almost by accident, with smiths not aware that carbon from coal fires became alloyed in the iron they were working. Because the chemistry of this was not known, and steel seemed to "appear" at random from working iron in a forge, steel was an almost magic metal. Indeed, several famous steel swords throughout history were actually given names because of their abilities or perceived abilities. Arguably the most famous is Excalibur.

In more modern times, steel has been manufactured with as wide an array of different compositions as the just mentioned bronze, and once again ASTM has an entire set of codes for different grades of steel. Cities such as Pittsburgh, Pennsylvania, became known for their steel output, especially as it met the needs of the military during the Second World War.

A third metal that has been sought out, worked, and alloyed for millennia is gold. One of only a few metals valued simply for its looks, gold has such a well-known ancient past that it is mentioned in virtually all the world's major religious books, and routinely found among grave goods in widely separated cultures throughout the world. Gold is often alloyed with copper because the resulting alloys are harder and more durable. The system used to classify gold by its purity – carats – is itself an ancient one. Pure gold is 24 carat, and grades such as 22-carat, 18-carat, 14-carat, and even 10-carat gold are still used in jewelry and other applications.

While many of us think of gold in terms of its use as a source of wealth, it also finds use in electronics and several other niche applications because of its superior conductivity [3]. But gold coins have been produced by nations for centuries, and still represent a means of storing wealth for many people in the world today, as shown in Figure 2.1. Those who store some portion of their wealth in gold purchase either coins or gold ingots of known weight.

Figure 2.1: Collectible gold coins.

2.1.2 Ceramics and glasses

The two materials, ceramics and glasses, are often treated together because both are created from sand and additives, both are heated to high temperatures in an oven, and both are brittle and can shatter once made into some usable object. The obvious difference between the two is that ceramics are almost always opaque, while glasses are generally transparent.

Flat glass is now made with float glass process, in which molten glass is cooled on a pool of molten tin metal. But glass can still be formed into bottle and other containers using specialized glassblowing apparatus, such as torches and furnaces. The end results, such as artistic bottles, as shown in Figure 2.2, can be useful and decorative.

Ceramics are thought of by the general public as a material from which containers and tableware are made, and indeed, many artisanal objects are still made of ceramics in a traditional manner. Figure 2.3 shows a vase that was made by turning clay on a wheel – the traditional means – then glazing and firing the resulting item.

Figure 2.2: Artisanal glass bottle.

But ceramics also find use as bricks, tiles, components in cements, and electrical insu-
lators (as has glass, in the past). Additionally, ceramic formulations have found uses
in bioimplants to the human body, and in some aerospace applications. Such will be
discussed in more detail in Chapter 15.

Figure 2.3: Ceramic vase made on a pottery wheel.

2.1.3 Plastics

The term "plastics" is one recognized by the public as one of a few materials used for a huge number of common objects, but the term used by the scientific community for this class of materials is more specifically "polymers." Even this term does not always encompass all of the materials that have the general properties of low density, ease of deformability, and low melting point that we consider normal for polymers, since what is generally called "rubber" ends up being excluded when such properties are used to define these materials.

Those plastics made on an industrial scale – the largest scale production plastics – all have a resin identification code, or RIC, to identify them. We will discuss this in more detail in Chapter 7. But a great many plastic materials have been produced with some end result in mind, often a specific flexibility, density, or durability.

Plastics that do not have a carbon-atom-based backbone also have been made in large amounts. Silicones are plastics made with a backbone of alternating silicon and oxygen atoms. What are called the side chains, the atoms attached to the silicon atoms which are not part of the material's backbone, often determine the bulk, macroscopic properties of a specific silicone. Figure 2.4 shows an example of the repeat units of a silicone molecule.

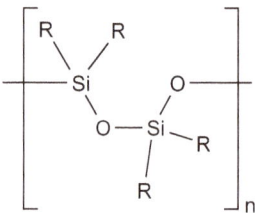

Figure 2.4: Lewis structure of the repeat unit of a silicone polymer.

The "R" groups shown in the Lewis structure of Figure 2.4 can vary widely. Two examples are a methyl group, $-CH_3$, or a phenyl group, $-C_6H_5$. Having such groups incorporated into silicones alters their stiffness, their durability, and their degradability, among other properties.

Another polymeric material sometimes discussed as a stand-alone material, and sometimes considered part of the greater subject of plastics is rubber. We will discuss this in more detail in Chapter 7, but will say here that the original rubber is extracted from the *Hevea brasiliensis* tree by cutting slices into the tree, and collecting the exudate sap. Figure 2.5 shows the repeat Lewis structure of the isoprene unit.

The Second World War was the single event that promoted the trial of a large number of synthetic rubber formulations in a very short time. Many have since been adopted for different uses. Much like silicones, changing the atoms that make up the side chains of rubber molecules changes the macroscopic properties of the resulting material.

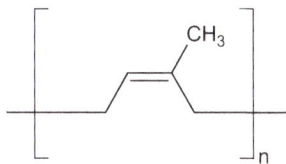

Figure 2.5: Lewis structure of the repeat unit of natural rubber.

2.2 Nonsolid phases of matter

There are certainly some materials that have been developed and have become increasingly useful in recent years which are not normally used in the solid state, and yet that still fall under the general subject of materials chemistry. For example, the idea of using sulfur hexafluoride gas (SF_6) as an insulator between panes of glass in household windows is certainly one that is becoming established and proven. Likewise, the field of ionic liquids – materials that are ionic, yet that are not solid at room temperature – has gone from a series of curiosities to a number of useful materials. Thus, materials chemistry may have the solid state as its heart, but now includes other materials in other states of matter. We briefly discuss a few of these, below.

2.2.1 Ionic liquids

Ionic materials are often thought of as simple solids, but the field of ionic liquids is one that has become widespread in the past three decades [4, 5]. These materials are ionic salts, but exist as liquids at ambient temperature and pressure. The idea of ambient temperature and pressure being essential to a material being called an ionic liquid is because many salts can be liquids at highly elevated temperatures. Sodium chloride, for example, is used in a molten state in what is called a Downs cell, a means of electrolytically separating sodium metal and chlorine gas.

Ionic liquids are sometimes referred to as "room-temperature ionic liquids." Figure 2.6 shows a common ionic liquid, 1-butyl-3-methylimidazolium hexafluorophosphate, sometimes abbreviated [BMIM]PF$_6$. Many others have been made as well.

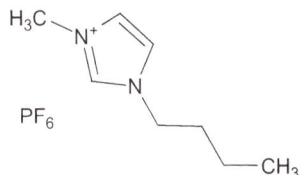

Figure 2.6: Lewis structure of [BMIM]PF$_6$.

Numerous applications for ionic liquids have been attempted and examined, since they possess the potential to act as electrolytes, and have uncommon abilities as solvents. Yet their adoption for some large-scale commercial use has thus far been limited to the production of gasoline components (gasoline is a mixture of isomers of octane) in a catalytic role.

2.2.2 Ferrofluids

The idea of a fluid that responds to a magnet has its origins in the 1960s, specifically in NASA. The interesting problem arose as to how to make a fluid flow in a specific direction while in microgravity. The answer to this challenge came to Steve Papell, who patented the idea of using fine iron particles in the fluid that needed to flow in a specific direction [6]. By applying magnets in the proper positions, a fluid could be made to move in a specific direction, even in microgravity. Today, ferrofluids can be made with a variety of particle sizes, and using a wide array of oils, one of which is shown in Figure 2.7. When the particles of iron are large enough, they can sometimes be separated from the surrounding oils.

Figure 2.7: Ferrofluid.

2.2.3 Novel gases – and their uses

While materials chemistry tends to focus on the solid state, as mentioned, there are some gases which have a place in the field as well.

The major industrial gases have been used for decades, meaning nitrogen, oxygen, argon, hydrogen, acetylene, and ethylene. The first three are separated from liquified air, hydrogen is stripped from methane, and the latter two have crude oil as their source. But there are several other gases that find their wide use as well. Not all of them are natural, or extracted from natural sources. Table 2.2 shows many of them, but is not an exhaustive list. Note that on this list, any gas that contains one or more fluorine atoms is a man-made synthetic gas. Essentially all naturally occurring fluorine is in the mineral fluorite, or CaF_2.

Table 2.2: Gases produced on a large scale.

Name	Formula	Source(s)	Example use
Acetylene (ethyne)	C_2H_2	Controlled methane combustion	Welding
Ammonia	NH_3	Air and methane	Fertilizer
Butane	C_4H_{10}	Crude oil	Fuel for lighters
Butene	C_4H_8	Hydrocarbon feedstock, catalytic cracking	Fuel additive
Carbon dioxide	CO_2	Air distillation, hydrocarbon combustion	Refrigerant
Carbon monoxide	CO	Coal (traditionally)	Industrial feedstock
Ethane	C_2H_6	Natural gas	Precursor to ethylene
Ethylene (ethene)	C_2H_4	Ethane	Polyethylene monomer
Helium	He	Natural gas wells	Cryogen, as a liquid – can be superconducting
Hydrogen chloride	HCl	Direct combination of H_2 and Cl_2, or as part of organofluoride production	Vinyl chloride production
Methane	CH_4	Natural gas wells	H_2 production, fuel
Neon	Ne	Liquified air	Vacuum tubes
Nitrous oxide	N_2O	Ammonium nitrate	Surgical and dental anesthetic
Nitrogen trifluoride	NF_3	Ammonia and fluorine gas	Microelectronics manufacture
Propane	C_3H_8	Crude oil or natural gas	Fuel
Propylene (propene)	H_2CCHCH_3	Steam cracking of hydrocarbon feeds	Polypropylene monomer
Sulfur dioxide	SO_2	Combustion of sulfur	Food preservative
Sulfur hexafluoride	SF_6	Sulfur and fluorine gas	Electric insulator

It is worth noting that the histories of several of these gases, such as hydrogen chloride or sulfur hexafluoride, are not really that old. Likewise, the hydrocarbons that are listed generally have seen increased use as the production of crude oil has increased with time. In the past few decades, however, the importance of all of these has risen, applications for them have increased, and they are now materials that are needed for more than just the examples noted in Table 2.2.

2.3 Recycling or downcycling

The design of materials beyond the traditional materials of wood, metal, glass, clays, and leather has a history that is generally a century old, and that originally took no notice of what would become of the material or user end item after its intended, usable life span was completed. Indeed, after the Second World War, numerous new materials, many of them plastics, were advertised as being designed to last "forever." When such materials were first made, and when consumer end use items were made from them, it appears that no one took into account the idea that having a material last "forever," and building an economy of constant consumerism were two forces that were in opposition to each other.

In the United States, the idea of recycling items that have been used and discarded by some person caught on over the course of time, with large and noticeable strides being made as the 1960s ended and the 1970s began. It is difficult to pin the reason for this to a single cause, but the publication of Rachel Carson's book, *Silent Spring* appears to have had much to do with it [7]. While the book does not directly treat the idea of recycling, its treatment of how the natural world was being poisoned by the widespread, unregulated use of synthetic pesticides dovetailed well into the greater environmental movement which was aimed at cleaning and remediating our soils, waterways, and air [8]. It is now quite common to see recycling bins and trash bins side by side in stores and businesses, and an example of which is shown in Figure 2.8. As well, what are called curbside recycling programs, in which home owners recycle many materials regularly, are now very common. Figure 2.9 shows the recycling bins for one of what must now be thousands of such programs throughout the United States. Canada, Europe, Australia, and many other countries and regions have such programs in place.

It may seem obvious that it is easier to clean up or recycle solid materials, such as glass, plastic, metal, and paper, than it is to recycle liquids or gases. This being said, motor oils are now often recycled at garages, by allowing them to settle in large drums, then reusing the lightest, cleanest portion.

As well, while it is easier to recycle or reuse solids and liquids than gases, the solution to gas-based pollution tends to be capture of gases and some form of sequestration of gases that are causing pollution. One extreme example is the air in London in 1952. Now referred to as the "Great Smog of London," this was a brief period

Figure 2.8: Recycling bins.

Figure 2.9: Household recycling bin.

in December 1952 in which the soot and smog in the city were so dense that they appear to have caused over 4,000 fatalities – and which ultimately brought about the Clean Air Act, passed by Parliament in 1956. As a second example, in and near the city of Pittsburgh, and the surrounding Three Rivers valley region, in the early 1960s, people driving into the city needed to have their headlights on at any time of the day, as well as night, simply because the smog and smoke from local steel mills was so thick that it was difficult to drive otherwise. By the 1970s, when the area steel manufacturers had ensured that their mills captured exhaust gases and soot, excluding carbon dioxide, the region's air became clean enough again that air quality and visibility was not a problem.

References

[1]   G. Hu, X. Cai, Y. Rong. Materials Science, Walter DeGruyter, GmbH, ISBN: 978-311049534-8.
[2]   M.A. Benvenuto. Metals and Alloys, Industrial Applications, 2016, Walter DeGruyter, GmbH. ISBN: 978-311040784-6.
[3]   U.S.G.S. Mineral Commodity Summaries, 2024, downloadable, as a pdf.
[4]   R. Fehrmann, C. Santini. eds. Ionic Liquids: Syntheses, Properties, Technologies, and Applications, Walter DeGruyter, GmbH. ISBN: 978-311058363-2.
[5]   W. Lee, S. Kumar. Unconventional Liquid Crystals and Their Applications, 2021, Walter DeGruyter, GmbH. ISBN: 978-311058437-0.
[6]   US Patent 3215572A Solomon Steven Papell. Low Viscosity Magnetic Fluid Obtained by the Colloidal Suspension of Magnetic Particles, 1963.
[7]   R. Carson. *Silent Spring*, 1962.
[8]   A. Alexander, S. Pacucci, F. Charnley. DeGruyter Handbook of the Circular Economy, 2022, Walter DeGruyter, GmbH. ISBN: 978-311072337-3.

Chapter 3
Synthetic chemistry

3.1 The search for novel materials

As mentioned in Chapter 1, there has been an enormous surge in the search for and the invention of synthetic materials in the last seven decades, a significant amount of it is originally driven by the needs of several nations during the Second World War. The availability of inexpensive crude oil at the end of the nineteenth century and the beginning of the twentieth, and the necessities of large-scale production of products from oil, as well as the metals, such as steel, copper, lead, and brass just before and during the Second World War, meant that materials which had been used traditionally to produce some end product, but that were now in short supply, had to be supplemented with new materials that performed at the same level and in the same manner as the old. The pace of research at the time of that war was faster than anything that had been seen before. And the end results are, among other things, a series of developments and materials that are still with us today [1]. One example out of a multitude is synthetic rubber versus natural rubber. Another is nylon versus the traditional silk. Table 3.1 shows a nonexhaustive list of such materials.

Table 3.1: Traditional materials and modern counterparts.

Traditional material	Example use	Modern substitute/ replacement	Comments
Ivory	Ornamental items	Cellulose nitrate	
Leather	Furniture, footwear	Naugahyde	First produced, 1914
Rubber	Raincoats	Gortex® fiber	Gortex first produced, 1972
Silk	Stockings, parachutes	Nylon	First produced, 1939
Wool, cotton	Clothing	Acrylic	First produced, 1944

The term "synthetic" is another that a person can argue has fuzzy edges. For example, most chemists would claim that the production of iron, its roots lost in antiquity, is not synthetic, precisely because it is so established, having seen evolutionary but not revolutionary improvements over the past two millennia. Yet iron must be extracted from its ores through the addition of carbon as a reducing agent, and thus can be considered a form of synthesis. Figure 3.1 shows the general reaction chemistry for this.

The manipulation of the olefinic bond, or double bond, on the other hand, to produce a wide variety of small molecules as well as polymers never seen prior to the advent generally of the twentieth century is a field that all would claim results in syn-

https://doi.org/10.1515/9783112205822-003

thetic materials. Figure 3.2 shows an example of the production of 1,2-dichloroethane from ethylene.

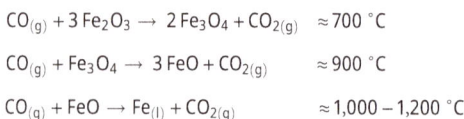

$$CO_{(g)} + 3\,Fe_2O_3 \rightarrow 2\,Fe_3O_4 + CO_{2(g)} \quad \approx 700\ ^\circ C$$

$$CO_{(g)} + Fe_3O_4 \rightarrow 3\,FeO + CO_{2(g)} \quad \approx 900\ ^\circ C$$

$$CO_{(g)} + FeO \rightarrow Fe_{(l)} + CO_{2(g)} \quad \approx 1{,}000 - 1{,}200\ ^\circ C$$

Figure 3.1: Production of iron from ore.

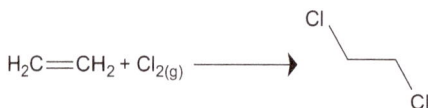

$$H_2C{=}CH_2 + Cl_{2(g)} \longrightarrow$$

Figure 3.2: Production of chlorinated ethane from ethylene.

A further example of a synthetic chemistry, straddling some line between these two examples of iron and double bond chemistry, is the production of leather, which also has roots lost in history. But as time progressed and knowledge grew, the production of leather became less a cottage industry and more an established process, one that requires several chemical transformations to cure animal skins, strengthen them, and prevent them from rotting.

Although we devote chapters to polymers, metals, and ceramics, we will look at them here, briefly, in terms of their source materials, and their synthesis and production.

3.2 Organic synthesis

A great deal of synthetic chemistry is essentially organic chemistry, the transformation of one or more organic molecules into some new material, often some molecule, that is more complex than the starting material or materials. This being said, transformations can also result in molecules and materials that are less complex than the chemicals from which they started. And while it may seem to be even the most ardent of organic chemists that the definition of organic chemistry is something like "the chemistry of carbon and carbon compounds," there are even gray zones in a definition as simple as this. For example, do chemical reactions involving some transformation of the carbonate anion (the CO_3^{2-} anion) qualify as being organic? Do reactions involving or producing carbon black, aka soot, qualify as organic chemistry?

For the purposes of the greater discussion of materials chemistry, in this chapter and later in this book we will concern ourselves less with some all-inclusive definition of organic chemistry and organic synthesis, and focus on chemical transformations that produce or utilize organic chemicals.

3.2.1 Polymers

The search for and production of novel plastics and polymers appears to be a never-ending one. There are currently six plastics produced on such a large industrial scale that they dominate the markets and that are seen by the consumer throughout the world. They are listed in Table 3.2.

Table 3.2: Six major plastics.

Name	Abbreviation	Common uses	RIC code	Comments
Polyethylene terephthalate	PET or PETE	Bottles, fibers	1	Often recycled
High-density polyethylene	HDPE	Bottles, jugs, plastic lumber	2	
Polyvinyl chloride	PVC or V	Piping, construction materials	3	
Low-density polyethylene	LDPE	Plastic bags, containers, tubing	4	Not usually recycled
Polypropylene	PP	Fibers, automotive parts	5	
Polystyrene	PS	Plastic utensils, packing "peanuts"	6	Usually very low density, not often recycled

Polymers are generally considered to be high-performance materials that are low in density when compared to most metals, and high in deformability when compared to ceramics and glasses. Their low density makes their light weight a useful aspect of any items made from plastics, such as bottles and other containers, as shown in Figure 3.3. The combination of low density and enhanced durability has made composite plastics – materials made from more than one starting polymer – a useful choice for several applications. There are numerous examples of this, including what might seem like an odd one – the stock of the military M16 rifle. Previous military rifles were made with wooden stocks. The choice to switch to a composite plastic was because it makes the weapon lighter.

Plastics are also widely used in applications where flexibility is an important feature. Using the medical profession as one very large example, it is hard to imagine a hospital, doctor's office, or dental office functioning without a wide array of flexible, plastic items and equipment.

All six of the polymers listed in Table 3.2 utilize some fraction of crude oil as their feedstock, or some starting materials made from crude oil. Five of the six of them depend upon the controlled opening of a double bond to affect polymerization from the monomer to the polymer, the sixth one (polyethylene terephthalate, PETE) being a

Figure 3.3: Example of plastic containers.

condensation copolymer utilizing two starting materials. The older term "olefin" is still used in industry for any material for which the double bond is the functional group that is active and of interest [2]. Figure 3.4 shows the idealized chemistry for the production of polyethylene.

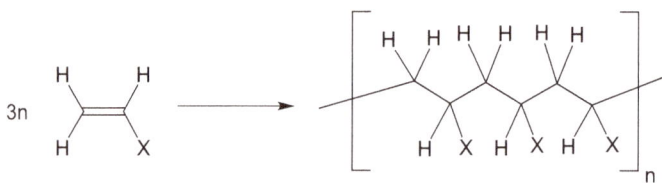

Figure 3.4: Polymerization of olefins.

As mentioned, PETE does not utilize the polymerization of a single monomer but depends on the controlled mixing of two components in as precise a one-to-one ratio as possible.

The RIC column of Table 3.2 stands for "Resin Identification Code," and is a means whereby these plastics can be identified for recycling, an industry that is now quite mature. For those plastics which are not one of these six, the RIC "7" is routinely used.

3.2.2 Organic synthesis and the carbon–halide bond

A large number of novel fine chemicals have been created by the synthesis of two or more organic molecules that are precursors to some final product. In many cases, one of these precursor molecules is some form of organohalides. As illustrated in Figure 3.5, the carbon–halogen bond of carbon and chlorine, carbon and bromine, as well as carbon and iodine is sufficiently reactive that it can be broken under conditions available to the synthetic chemist or chemical engineer, allowing the organic fragment to bond to some other organic molecule.

$R_3C—Cl$ and $R_3C—Br$ and $R_3C—I$

Figure 3.5: The halide bond.

Note that the carbon–fluorine bond is not included in those halide bonds just listed. This is because its bond energy is strong enough that organofluorides are not easily or routinely used as components in organic syntheses. Indeed, a fluorine atom can be utilized in a specific spot in a complex molecule where a hydrogen atom would normally be precise because it decreases the reactivity at that site.

Since the organohalide bond is important in the designed organic synthesis, and of the three, the organochlorine bond is probably the most widely known and used, we should take a moment to determine how such bonds are themselves made. Ultimately, we should determine the source materials, including chlorine. The organic component of any organohalide can be derived from adding a hydrogen halide to a double bond, as shown in Figure 3.6. But alcohols are often converted to their alkyl halide form as well, as shown in Figure 3.7.

Figure 3.6: Alkyl halide formation from double bond.

Figure 3.7: Alkyl halide formation from alcohol.

It is also possible, starting with some alkanes, to produce alkyl halides via direct addition of the halide or hydrogen halide to the alkane.

Since alkyl chlorides are so common, it is important that we know where the chlorine comes from. An enormous amount of the chlorine in the world exists in the form of sodium chloride, some in mineral deposits, much in the oceans, or brine lakes. Elemental chlorine can be produced by the electrolytic separation of molten sodium chloride at high temperature in what is called a Downs cell, or from the electrolysis of aqueous brine solutions. The chemistry of the latter, which is called the Chlor-Alkali process, is shown in Figure 3.8. This is a chemical reaction that is one of the most common runs in the industry.

$$2\,NaCl_{(aq)} + H_2O_{(l)} \longrightarrow 2\,NaOH_{(aq)} + 2\,H_{2(g)} + Cl_{2(g)}$$

Figure 3.8: Chlor-Alkali process.

The Chlor-Alkali process has for decades been used to produce sodium hydroxide as the product that drives the reaction in terms of economics. But the production of chlorine through it is also important, as is the elemental hydrogen.

3.2.3 Organic synthesis and the double bond

The double bond, still often referred to as the olefinic bond, is another central aspect to organic synthesis. As shown in Figures 3.6 and 3.7, addition can be of some substance to both ends of the bond. But a great deal of effort and design has been put into the development of methods that direct synthesis to either the cis-side of the double bond – both atoms to the same side – or to trans-addition to the double bond. Figure 3.9 shows idealized versions of this.

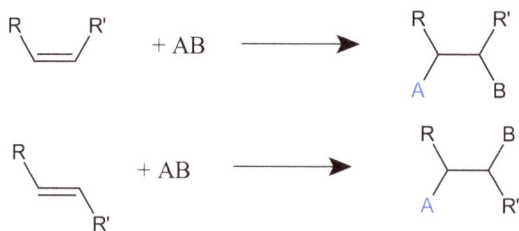

Figure 3.9: Cis- or trans-addition to a double bond.

Many organic chemistry textbooks illustrate these selective additions using some ring structure as a starting point, since the hindrance of the ring tends to favor trans-additions to the double bond. But creative and often elegant solutions have been found to direct an addition to either a cis- or a trans-arrangement. This is designed to produce a material with characteristics favorable to a specific application [3, 4].

Figure 3.4 shows the basic reaction chemistry of an olefin polymerization. Early patents for the synthesis of polyethylene – the simplest of polyolefins – are now over 60 years old [5]. Since these first patents were issued, numerous others have been filed and issued, because the types of monomers, additives, and catalysts have been developed into a wide suite of possibilities [6, 7].

3.3 Inorganic syntheses

What can be called the order inherent predictability in syntheses that are based on carbon – the usefulness of the carbon-halide bond, for example – does not exist in syntheses that are inorganic. This is because there are many more elements from which synthetic components are sourced, and also because numerous inorganic syntheses depend on rearrangements of ionic bonds, or bonds that are much more ionic in character than they are covalent in character. The production of metals or metal alloys and the production of ceramics and glasses represent two broad areas of inorganic synthesis that require wildly different approaches to produce useful, novel materials.

3.3.1 Metals

It may seem odd to think of metals as new materials, or as novel forms of matter. This may be because several metals have been known since ancient times – including gold, silver, copper, iron, lead, tin, and mercury – and their production and uses today are well established. Iron, for example, has been worked for millennia, although the number of ores from which it can be obtained, as shown in Table 3.3, has not been known for that long.

The chemistry by which iron is reduced from an ore to the elemental metal – the means by which it is synthesized, as it were – has been shown, but is presented again in Figure 3.10 with a further required starting material illustrated as the first reaction. The presence of carbon as a reducing agent is essential to the refinement of iron. Still, these are general steps and may vary depending upon the starting ore.

It is noteworthy that in each step, the carbon starting material has been oxidized to carbon monoxide, which functions as a reducing agent, being itself further oxidized to carbon dioxide. The carbon source for carbon monoxide is coke, which is added to the blast furnace in which the iron ore is heated. The coke is combusted with blasts of air, partially oxidizing it. Finally, note that the product, iron, is a liquid. The molten iron can be poured into forms to make what are still called pigs, which can then be flattened into sheets of varying thickness. From here, a wide variety of steels can be formulated.

Table 3.3: Iron ores.

Mineral name	Chemical formula	Iron, percentage	World location	Additional metals in ore
Ankerite	$Ca(Mg,Mn,Fe)(CO_3)_2$	Varies	Peru, Austria	Magnesium, manganese
Goethite	$FeO(OH)$	62.8	Australia, Brazil, Russia, South Africa	
Greenalite	$Fe_4Si_2O_5(OH)_4$	52.3	USA, Minnesota	Both Fe^{2+} and Fe^{3+}
Grunerite	$Fe_7Si_8O_{22}(OH)_2$	39.1	South Africa	
Hematite	Fe_2O_3	69.9	Australia, Brazil, Canada, China, India, Russia, South Africa, Ukraine, USA	
Laterite	Mixed $Fe_xAl_yO_z$	Varies	India, Australia	Aluminum, nickel
Minnesotaite	$(Fe,Mg)_3Si_4O_{10}(OH)_2$	30.7	USA, Minnesota	
Taconite	Fe_3O_4 mixed with quartz	*Often >15	USA, Minnesota, Michigan	

*Taconite routinely contains iron as dispersed magnetite particles.

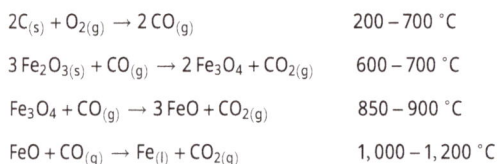

$$2C_{(s)} + O_{2(g)} \rightarrow 2\,CO_{(g)} \qquad 200-700\ °C$$

$$3\,Fe_2O_{3(s)} + CO_{(g)} \rightarrow 2\,Fe_3O_4 + CO_{2(g)} \qquad 600-700\ °C$$

$$Fe_3O_4 + CO_{(g)} \rightarrow 3\,FeO + CO_{2(g)} \qquad 850-900\ °C$$

$$FeO + CO_{(g)} \rightarrow Fe_{(l)} + CO_{2(g)} \qquad 1,000-1,200\ °C$$

Figure 3.10: Iron reduction and production.

Another metal known from ancient times but which has numerous modern applications is silver. When found as sulfide minerals, it is first concentrated, then refined. The Washoe process for this silver extraction was used while the Comstock Lode of Nevada was being mined, but has been displaced by the Reese River process. Very broadly, this involves the following nonstoichiometric steps:

1. Ore is pulverized to the size of grains of sand.
2. Mercury is added to make an amalgam.
3. Copper sulfate (bluestone) and sodium chloride are added.
4. Silver chlorides are concentrated in amalgamation pans.
5. Separation of impurities, leaving the amalgam.

It is the presence of mercury in any such process which is an environmental concern. Currently, most silver is refined via smelting and leaching processes. Electrolytic refining is often the final step. But the value of silver is such that improvements of any process are constantly examined.

The wide suite of uses for silver is dominated by electrical and electronics applications but also includes such things as pharmaceuticals and antimicrobial bandages [9].

Several metallic elements do not have long histories, though, having only been discovered in the late eighteenth or early nineteenth centuries. Thus, their properties, or more broadly the properties of their alloys, have not been completely determined or even studied. Many alloys have been created in the relatively recent past that are indeed new, and that have revolutionized parts of daily life. Some examples include the extensive number of marine brasses, so called because they are resistant to oxidation and corrosion in saline waters. Also, niobium–tantalum alloy wire, and "supermagnets," or neodymium magnets are relatively new alloys with niche applications. Also, the large-scale use of aluminum for aircraft seems well established to us today, but is really only a use that came of age after the end of the First World War – since airplanes during that conflict were made largely of wood and canvas, aside from the engines. The Hall–Heroult process for the production of aluminum was only patented in the 1880s [10]. Additionally, there are also many forms of low-melting alloys which have found specific, sometimes niche, uses. Table 3.4 is a sampling, a nonexhaustive list, of modern alloys.

Table 3.4: Examples of modern alloys.

Name	Composition	First report, uses
Aluminum alloy 7075	1.2–1.6% Cu, 2.1–2.5% Mg, 5.6–6.1% Zn (and <½%: Cr, Fe, Mn, Si, Ti)	Aircraft
Aluminum alloy 7085		Aircraft/aerospace
Amalgam	45–55% Hg, and Ag, Sn, Cu, Zn	Dental fillings
Cupro-nickel	75% Cu, 25% Ni	"Silver" coins
Galinstan	68.5% Ga, 21.5% In, 10.0% Sn	Low melting alloys (freezes, −19 °C)
Magnox	Mg, Al	Nuclear reactor fuel rod sleeves
Monel®	66% Ni, 31.5% Cu, plus	
Muntz metal	60% Cu, 40% Zn, trace Fe	1832, ship hull linings
Neodymium magnets	Nd–Fe–B	Cell phones, first created in 1982 [8] Stereos, computers, TVs

Table 3.4 (continued)

Name	Composition	First report, uses
Nichrome	80% Ni, 20% Cr	Heating elements
Niobium–tantalum wire	Niobium and tantalum	Magnetic fields for NMR and MRI instruments. Specific alloy compositions can be proprietary [9]
Stainless steel		1913, included 12.8% Cr. [9]
Wood's metal	50% Bi, 26.7% Pb, 13.3% Sn, 10.0% Cd	Low melting applications

3.3.2 Ceramics

Somewhat like metals, it can be difficult to believe that ceramics qualify as new, synthetic materials, since they have been known since ancient times. But this ancient art of creating items from natural clays – still seen in the pottery of countless archaeological digs and among modern, artisanal potters – has evolved into a science of producing specific ceramics with exact, well-defined characteristics.

Unlike organic syntheses, ceramic formulations remain a matter of trial and error (very much like the creation of metal alloys). This may sound primitive or underdeveloped, but historically, before an understanding of the periodic table existed, the only way to produce a specific ceramic was to know the formula and reproduce it. The idea of systematically changing one component in the chemical composition of a ceramic to determine what the properties of the end product will be is a relatively recent development.

A dual example of ceramic synthesis that is well established, yet that started through trial and error, is two ceramic formulas that now qualify as classic: Wedgewood china and Meissen porcelain. Both have become well established as a high-end or luxury type of item, certainly in terms of consumer products such as dinnerware. But each company produces more than just this.

Figure 3.11 shows a Wedgewood box, made as an ornamental piece. Josiah Wedgewood founded the company that bears his name in 1759, and what is called Jasperware® has become famous worldwide. The blue color, which became essentially a trademark for the company, has been expanded to a pink-, a green-, and a black-colored ceramic of the same texture and other properties. The Wedgewood formula remains proprietary, although it is known that barium sulfate is more than 50% of the mixture, and that other barium salts, such as barium carbonate, are present, or have been in some formulations. Also, despite the confidential nature of this formula (and that of many others in other companies), the success of blue Jasperware has spawned a number of imitators, all of which are able to produce a ceramic material that has a similar blue color. As well, the use of ceramics has become so widespread that several societies exist to advocate for the use of such materials [11–14].

Figure 3.11: Porcelain Wedgewood box.

As far as ceramic formulas, Wedgewood now markets far more products than just the different color Jasperware, all of them ceramic based. Josiah Wedgewood's early, systematic search for a desired ceramic product is arguably one of the first scientific searches for a marketable material that has been conducted in history.

Less known in the English-speaking world is the ceramics of Meissen porcelain. The first such porcelain, produced under the direction of Johann Friedrich Boettger, had a dark red color – and remains a hallmark of Meissen porcelain today. Figure 3.12 shows an example of a medal made in Meissen porcelain.

Figure 3.12: Meissen porcelain medal.

Meissen porcelain was produced as early as 1710, and like Wedgewood keeps its formulas proprietary. But the dark red color that has come to characterize this porcelain is because the raw material for the finished porcelain is of this color, and is found where Boettger established his business in Germany.

The steps involved in ceramic production can be very specific, depending on the product being made. There are, however, some basic, generic steps almost all the production of ceramics. They include:

1. Enhancement – sometimes called beneficiation.
2. Mixing and forming – to ensure even blending of all components
3. Green machining – the machining and forming of the material before heating
4. Sintering – heating and increasing pressure
5. Drying and thermal processing
6. Glazing – not needed in some applications
7. Firing – the gradual increase in heat of the material to finalize the form.

Because there are many types of ceramic materials, there can be significant variation in the details of each of these steps.

3.4 Hard materials – the need by NASA

Numerous applications exist for hard materials, not all of them related to NASA and space travel. We use this as a starting point, though, because launching any object or ship into space, whether manned or unmanned, requires materials that are both hard and lightweight. The development of such materials has been one of the many challenges NASA has undertaken since the beginnings of a US space program just over 50 years ago.

3.4.1 Materials critical to the space aerospace industry

Aluminum is one of a select few metals that have seen extensive use as a hard material in the aerospace industry. It has low density – only 2.7 g/cm^3 – which is a property the general public associates with aluminum, but it is also strong and able to deform without breaking. These are important properties when a manned or unmanned object must be pushed through the atmosphere to space. Aluminum production is always via the Hall–Heroult process and will be discussed in more detail in Chapter 8.

Titanium is another metal needed both by NASA and in the larger aerospace industry. The general public tends to associate titanium with high strength but it is also of relatively low density – only 4.5 g/cm^3. As a point of comparison to these two metals, iron has a density of 7.87 g/cm^3. Once again, this becomes critically important when any object is lifted into space. Titanium is routinely produced via the Kroll process, which is a reduction from titanium tetrachloride, as shown in Figure 3.13. It is evident that significant amounts of chlorine are also required for this production.

$$2\,Mg + TiCl_4 \longrightarrow 2\,MgCl_2 + Ti_{(s)}$$

Figure 3.13: Basic chemistry of the Kroll process.

Modern spacecraft as well as airplanes rely more on ceramic tiling and surfaces than they have in the past. Such tiles absolutely must be heat resistant, and shield the craft as it leaves or reenters the Earth's atmosphere. The space shuttles have routinely used ceramics made from silica fibers. Such fibers are made by heating silica, often with a proprietary blend of other materials which can include aluminum oxide or with fluorides, to more than 2,000 °C, often in a microwave furnace.

3.4.2 Diamond

Diamond is an allotrope of carbon (C_{10}) that possesses a face-centered cubic arrangement, as shown in Figure 3.14. Natural diamonds are mined in a relatively few places on the Earth. Synthetic diamonds are produced by heating and applying extreme pressure to graphite, the much more common allotrope of carbon (C_6).

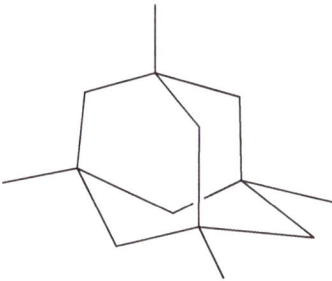

Figure 3.14: Unit structure of diamond.

The hardness of diamonds is well known, and to the general public, they are usually known as the hardest materials on the Earth. It is also well known that large diamonds are extremely expensive, and that manufacturing diamonds of any size requires enormous pressure. These latter two factors have been the reason that limits diamond's uses as a large-scale material.

Synthetic diamonds have traditionally been smaller than those suitable for use as jewelry. Recent advances, however, have improved the quality and size of synthetic diamonds. These include the ability to create extreme high pressures and temperatures and advances in chemical vapor deposition.

3.4.3 Nitrides and carbides

Boron nitrides (BNs): There are several forms of BN, with what is called c-BN being the cubic form. This c-BN is the diamond analog and is made by heating hexagonal-BN at extreme pressures, much like diamond is produced from some graphite starting material. BNs are not valued as gems in the same manner as diamonds, and most BNs find use as industrial abrasives. As well, there are several other nitrides – metal nitrides – that find their use as hard materials.

Silicon carbide, aka carborundum, is made by the high-temperature reaction of silica with carbon, in a furnace. The naturally occurring form is rare and is called moissanite, after Henri Moissan, its discoverer.

Sometimes called the "poor man's diamond," moissanite is produced by the combination of its two elements under extreme heat and pressure. Small, precut crystals are routinely used as seeds in producing larger moissanite crystals.

References

[1]   V.M. De Franca Bezerra. Chemical Process Synthesis: Connecting Chemical with Systems Engineering Procedures, 2022, Walter DeGruyter GmbH. ISBN: 978-311046825-0.

[2]   M. Elzagheid. Macromolecular Chemistry: Natural and Synthetic Polymers, 2022, Walter DeGruyter GmbH. ISBN: 978-311076276-1.

[3]   K. Maly. Synthesis of Aromatic Compounds, 2022, Walter DeGruyter GmbH. ISBN: 978-311056267-5.

[4]   N.P.S. Chuahuan, N.S. Chundawat, D. Singh. Organometallic Reagents in Organic Synthesis, 2022, Walter DeGruyter GmbH. ISBN: 978-150151916-1.

[5]   US Patent: US2907805. H. Bestian, E. Prinz. Process for the Preparation of Liquid Ethylene Polymers, 1959.

[6]   US Patent: US3257332A. K. Ziegler, B. Heinz, M. Heinz. *Polymerization of Ethylene*, 1966.

[7]   US Patent: US9090719B2. J. Berthold, B.L. Marczinke, D. Doetsch, I. Vittorias, D. Lilge, H. Vogt, J.-G. Mueller. *Polyethylene Composition and Finished Products Made Thereof*, 2015.

[8]   M.A. Benvenuto. Metals and Alloys, Industrial Applications, 2016, Walter DeGruyter, GmbH. ISBN: 978-311040784-6.

[9]   U.S.G.S. Mineral Commodity Summaries, 2024, Downloadable as: pubs.usgs.gov/publication/mcs2024.

[10] U.S. Patent. C.M. Hall, Manufacture of Aluminum. US400665A, 1889.

[11] American Ceramics Society. Website. (Accessed 8 June 2025, as: https://ceramics.org).

[12] United States Advanced Ceramics Association. Website. (Accessed 8 June 2025, as: https://advancedceramics.org).

[13] English Ceramics Circle. Website. (Accessed 8 June 2025, as: https://www.englishceramiccircle.org.uk).

[14] European Ceramic Society. Website. (Accessed 8 June 2025, as: https://www.ecers.org/).

Chapter 4
Chemical bonding

4.1 Ionic and molecular bonding

As early as any general chemistry course, students are introduced to the difference between covalent and ionic bonding in materials – the sharing of electrons versus the loss and gain of electrons, respectively. In doing so, this often gives the false idea that covalent bonding is reserved or localized within organic molecules, and ionic bonding is solely concentrated in inorganic materials – usually between a metal and a nonmetal atom – and that there is no "middle ground" in which a molecule may exhibit some covalent and some ionic characters. With further study, students are made aware that there are generally three types of bonding: covalent, ionic, and metallic, as shown in Figure 4.1. This is sometimes called a Ketelaar triangle, or a van Arkel–Ketelaar triangle, named after Jan Arnold Albert Ketelaar and Anton Eduard van Arkel, who did pioneering work on bonding types in the 1940s.

Figure 4.1: van Arkel–Ketelaar triangle.

Even on a qualitative level, the triangle makes it easy to see that any bond can have a certain amount of ionic character as well as covalent character, as well as metallic character. Using simple examples, a molecule of elemental chlorine, Cl_2, can be said to be entirely covalent. A unit of sodium chloride, $NaCl$, is generally considered completely ionic in nature. And an iron–nickel alloy, in which an atom of iron is bonded to one of nickel is considered entirely metallic. But more quantitative assessments can be made.

Measuring and determining how much covalent, ionic, or metallic character a bond has can be through more than one means, but a comparison of electronegativities of each atom in the bond is one that is generally accepted. The Greek letter χ – pronounced chi ("kai") – is used to represent electronegativity. The general values for elemental electronegativity range from a minimum of 0 to a maximum of 4. Perhaps obviously, when two elements have 0 electronegativity *difference*, the bond is covalent, whereas the purely ionic bond will have the highest difference in electronegativity. The following expression is used to quantify the percent of ionic character in a bond:

$$\% \text{ ionic character} = 1 - \exp[-0.25(\chi_A - \chi_B)^2]$$

https://doi.org/10.1515/9783112205822-004

Since the reaction is always unity minus the remainder of the equation, the answer will always be a decimal value less than one. Multiplying this by 100 will give the answer in percent. It is noteworthy that a bond between two metal atoms, specifically two in the transition metal section, often results in a χ difference of zero. This results in the entire term that is subtracted from 1 being equal to 1. Thus, such metal–metal bonds have no ionic character.

Using two compounds as examples of this, we can determine the ionic character of cesium fluoride, and compare it with that of cesium iodide.

First:

$$Cs\ \chi = 0.7$$

$$F\ \chi = 4.1$$

The electronegativity difference of 3.4 computes to a percent ionic character of 0.945 or 94.5% ionic for cesium fluoride – two ions that are as far apart on the periodic table as possible.

In comparison, putting in values for cesium and iodide

$$Cs\ \chi = 0.7$$

$$I\ \chi = 2.2$$

The smaller electronegativity difference of 1.5 computes to a percent ionic character of 0.431 or 43.1% ionic character – for two ions that are as far apart as they can be for a single row of the periodic table.

These types of determinations of covalent or ionic character based on electronegativity values for each atom in a bond do have their limits, specifically in the d-block elements and f-block elements. As can be seen from the values in the table in Figure 4.2, the differences in electronegativity values for these elements are very small, and are often nonexistent. The lanthanides and actinides have a particularly small range of electronegativity differences. This is because of the shape of both the d-electrons and the f-electrons – their directionality. Both sets of electrons are directional enough that the areas they occupy do very little to shield the nucleus from any region of space at the outer edge of an atom. The p-electrons with their "dumbbell" shape are much more effective at shielding a nucleus from any influences at the outer edge of an atom. The spherical s-atoms are the most effective in this regard.

Finally, the electronegativities of the noble gases are not routinely given, since their reaction chemistry is very minimal, they have full shells of valence electrons, and under normal circumstances they do not form bonds.

H 2.1																	
Li 1.0	Be 1.5				C - atomic symbol							B 2.0	C 2.5	N 3.1	O 3.5	F 4.1	
Na 0.9	Mg 1.2				2.5 - electronegativity							Al 1.5	Si 1.8	P 2.1	S 2.4	Cl 2.9	
K 0.8	Ca 1.0	Sc 1.3	Ti 1.5	V 1.6	Cr 1.6	Mn 1.6	Fe 1.7	Co 1.7	Ni 1.8	Cu 1.8	Zn 1.7	Ga 1.8	Ge 2.0	As 2.2	Se 2.5	Br 2.8	
Rb 0.8	Sr 1.0	Y 1.2	Zr 1.4	Nb 1.5	Mo 1.3	Tc 1.4	Ru 1.4	Rh 1.5	Pd 1.4	Ag 1.4	Cd 1.5	In 1.5	Sn 1.7	Sb 1.8	Te 2.0	I 2.2	
Cs 0.7	Ba 0.9	La 1.1	Hf 1.3	Ta 1.4	W 1.4	Re 1.5	Os 1.5	Ir 1.6	Pt 1.5	Au 1.4	Hg 1.5	Tl 1.5	Pb 1.6	Bi 1.7	Po 1.8	At 2.0	
Fr 0.7	Ra 0.9	Ac 1.1															

lanthanides and actinides, all between 1.0 and 1.3

Figure 4.2: Periodic table and electronegativities.

4.1.1 Ionic nature

What is called the ionic nature of a bond is more than simply the electrostatic connection between a metallic and a nonmetallic element [1–3]. For example, potassium oxide, which sometimes goes by the common name, potash, has the formula K_2O, while osmium tetroxide, aka osmium(VIII) oxide, has the formula OsO_4. Both can be considered ionic solids, yet one is highly reactive (K_2O), while osmium tetroxide is actually a volatile solid. Clearly, the nature of their ionic bonds must have some differences. We have seen how to compute the percent ionic character of each, but this does not necessarily correspond to their reactivity.

Figure 4.3 shows the repeat unit structure of an ionic solid that has a 1:1 cation-to-anion ratio. When extended throughout two dimensions, where it can be seen that each atom adjoins four other, to three dimensions, it can be envisioned that each atom then has six nearest neighboring atoms.

It should be noted, however, that not all 1:1 ionic solids have the structure shown here, which can be called primitive cubic or simple cubic (which will be discussed more in Chapter 5). If the size of the cation and anion is sufficiently different from each other, one ion can be positioned as the center point for eight surrounding ions.

4.1.2 Covalent nature

Covalent bonding is taught to students as early as the general chemistry class, and extensively in the organic chemistry class, those usually taken during freshman and

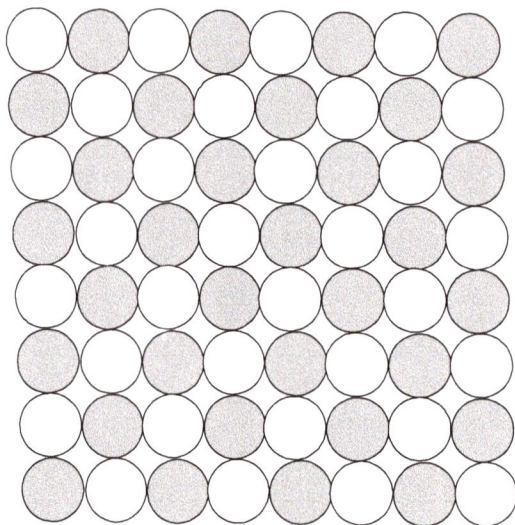

Figure 4.3: Repeat structure of a 1:1 MX ionic solid.

sophomore years in colleges and universities. The discussion often focuses on the hybridization of carbon atoms and sharing of valence electrons between them – as in Figures 4.4 and 4.5 – although covalent bonding certainly exists in inorganic substances as well. Silicones and elemental sulfur are two examples, as shown in Figures 4.6 and 4.7.

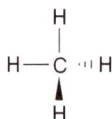

Figure 4.4: Covalent bonds in methane.

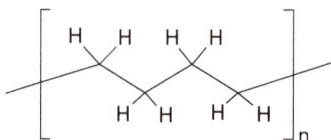

Figure 4.5: Polyethylene.

A major difference between covalent bonding and either ionic or metal bonding is that while the covalent bonding in a molecule is strong, this strength does not extend from one molecule to another (which we discuss in more detail, further).

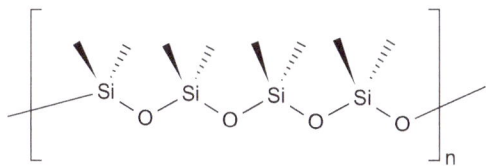

Figure 4.6: Example of a silicone.

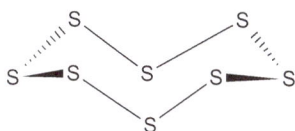

Figure 4.7: Elemental sulfur.

In Figure 4.7, note that what are called sulfur crowns have eight atoms in a truncated ring, each bonded covalently to those nearest to it (the shape resembling a crown). In Figure 4.6 it can be seen that silicones can be composed of very long-chain polymers with no carbon atoms in their backbone, rather having alternating silicon and oxygen atoms. Thus, they are molecules with large numbers of covalent bonds, but no carbon-to-carbon atom bonds in their main chains.

In general, covalently bonded molecules that are of low molecular weight tend to be gases at ambient pressure – in terms of hydrocarbons, up to butane, or C4. Middle range hydrocarbons tend to be liquids – such as octane. And higher molecular weight covalent compounds such as polymers, be they hydrocarbon-based or otherwise, tend to be solids at ambient temperature and pressure.

4.1.3 The dative bond or coordinate covalent bond

The field of coordination chemistry depends on what is sometimes called the dative bond, or the coordinate covalent bond. From the Latin "dativus," the term means "giving." In a dative bond, two electrons still are incorporated in a covalent bond, but both are supplied to the bond by a single atom. One atom is thus electron rich while the other is electron poor. Perhaps, the classic example of this is the bond between a nitrogen and a boron atom in some small molecules, such as Cl_3BNCl_3. While this does illustrate the idea of the electron-rich nitrogen atom providing two electrons from the lone pair on the nitrogen to a bond with an electron-poor boron atom, the chemical is so exotic that it is doubtful if many chemists or engineers have ever seen such a chemical.

A material that shows donor–acceptor bond which is much more common is in blood, or more specifically, the heme in the hemoglobin in blood. Figure 4.8 shows the heme in hemoglobin, leaving connecting groups simply labeled with "R," and emphasizes the electron donation that occurs when the four nitrogen atoms in the macrocycle donate electrons to the iron(II) ion in the heart of the heme.

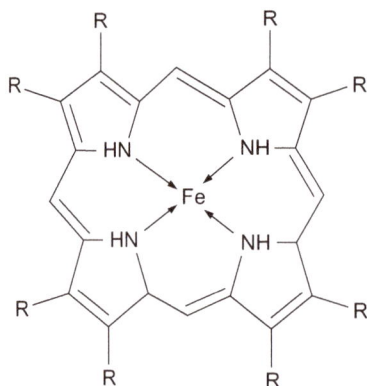

Figure 4.8: Lewis structure of a heme.

Another example is shown in Figure 4.9, the molecule salen. The name is shortened from **sal**icaldehyde and ethylenediamine (**en**), the two precursor molecules from which salen is always made. Here we emphasize that the Lewis base atoms in salen – the two oxygen atoms and the two nitrogen atoms – undergo coordinate covalent bonding when metal salts are exposed to the salen, usually in some polar, organic solvent.

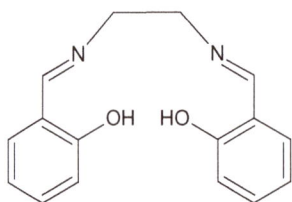

Figure 4.9: Lewis structure of salen.

4.2 Metallic bonding

Metals possess a different kind of attraction from atom to atom than do ionic solids or covalent materials [4, 5]. The metallic bond has in common with the covalent bond a sharing of electrons between atoms – but the metallic bond shares electrons with adjacent atoms in three dimensions. The model for this is often called the electron sea, because valence electrons move freely from one atom to another, and the nuclei can be considered "islands" which do not move. The metallic bond also has a wide range of affinities, or attractions. We can define it as one metal atom bonded to another of the same element, or as one metal atom bonded to another metal atom, but of a different element. Although such bonding can correlate in some way to a physical property, in some cases, a physical property can illustrate the differences in attractions. For example, titanium has a melting point of 1,668 °C, while mercury, a liquid at ambient temperature, has a melting point of −38.8 °C. A list of the melting points of the metallic elements is shown in Table 4.1. Note that the table has been arranged from lowest to highest melting point, and appears to have no correlation to atomic number.

Figure 4.10 shows in graph form the relationship between atomic number along the x-axis, starting with lithium, and melting point along the y-axis, in °C, of the metal elements. From it, we can see that there are highs and lows that can be considered periodic in nature. But direct correlations between atomic number (and, thus, atomic mass) and melting point that span the entire periodic table are difficult to find.

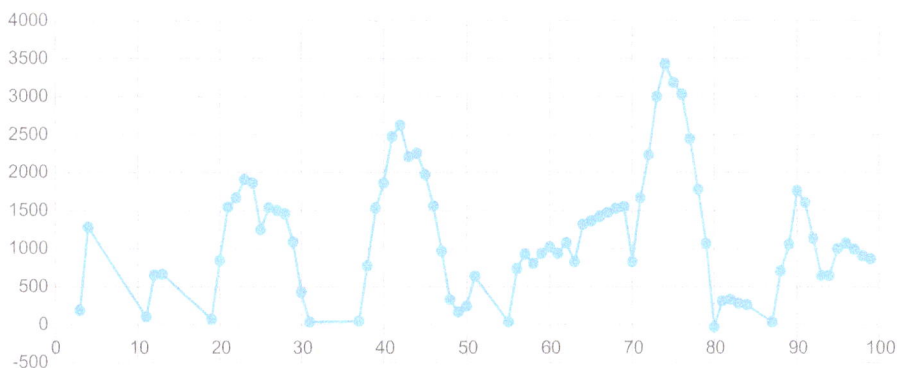

Figure 4.10: Melting points of the metal elements.

Like the metallic elements, some alloys have extremely high melting points, such as those containing tungsten. We can see from Table 4.1 that tungsten has the highest melting point of the metal elements. Alloys such as high-speed steel – an alloy containing tungsten – also have high melting points. Opposed to this, some alloys have extremely low melting points, such as galinstan, an alloy of gallium, indium, and tin, which melts at −19 °C.

Table 4.1: Melting points of the metal elements.

Name	Symbol	At. no.	Melting point (°C)
Mercury	Hg	80	−38.7
Francium	Fr	87	27
Cesium	Cs	55	28.6
Gallium	Ga	31	29.9
Rubidium	Rb	37	39.6
Potassium	K	19	63.4
Sodium	Na	11	98.0

Table 4.1 (continued)

Name	Symbol	At. no.	Melting point (°C)
Indium	In	49	156.8
Lithium	Li	3	180.7
Selenium	Se	34	221.0
Tin	Sn	50	232.0
Polonium	Po	84	254.0
Bismuth	Bi	83	271.5
Thallium	Tl	81	304.0
Cadmium	Cd	48	321.2
Lead	Pb	82	327.6
Zinc	Zn	30	419.7
Tellurium	Te	52	449.7
Antimony	Sb	51	630.9
Plutonium	Pu	94	640.0
Neptunium	Np	93	640.0
Magnesium	Mg	12	649.0
Aluminum	Al	13	660.3
Radium	Ra	88	700.0
Barium	Ba	56	729.0
Strontium	Sr	38	769.0
Cerium	Ce	58	798.0
Europium	Eu	63	822.0
Ytterbium	Yb	70	824.0
Calcium	Ca	20	839.0
Einsteinium	Es	99	860.0
Californium	Cf	98	900.0
Lanthanum	La	57	920.0
Promethium	Pm	61	931.0
Praseodymium	Pr	59	931.0
Germanium	Ge	32	937.4
Silver	Ag	47	961.0
Berkelium	Bk	97	986.0
Americium	Am	95	994.0
Neodymium	Nd	60	1,016.0
Actinium	Ac	89	1,050.0
Gold	Au	79	1,064.6
Curium	Cm	96	1,067.0
Samarium	Sm	62	1,072.0
Copper	Cu	29	1,084.6
Uranium	U	92	1,132.0
Manganese	Mn	25	1,244.0
Beryllium	Be	4	1,278.0
Gadolinium	Gd	64	1,312.0
Terbium	Tb	65	1,357.0
Dysprosium	Dy	66	1,412.0
Nickel	Ni	28	1,453.0
Holmium	Ho	67	1,470.0

Table 4.1 (continued)

Name	Symbol	At. no.	Melting point (°C)
Cobalt	Co	27	1,495.0
Erbium	Er	68	1,522.0
Yttrium	Y	39	1,526.0
Iron	Fe	26	1,535.0
Scandium	Sc	21	1,539.0
Thulium	Tm	69	1,545.0
Palladium	Pd	46	1,552.0
Titanium	Ti	22	1,660.0
Lutetium	Lu	71	1,663.0
Thorium	Th	90	1,755.0
Platinum	Pt	78	1,772.0
Protactinium	Pa	91	1,840.0
Zirconium	Zr	40	1,852.0
Chromium	Cr	24	1,857.0
Vanadium	V	23	1,902.0
Rhodium	Rh	45	1,966.0
Technetium	Tc	43	2,200.0
Hafnium	Hf	72	2,227.0
Ruthenium	Ru	44	2,250.0
Iridium	Ir	77	2,443.0
Niobium	Nb	41	2,468.0
Molybdenum	Mo	42	2,617.0
Tantalum	Ta	73	2,996.0
Osmium	Os	76	3,027.0
Rhenium	Re	75	3,180.0
Tungsten	W	74	3,422.0

Most scientists and engineers are probably not as accustomed to utilizing the boiling points of metals as they are the melting points. Still, certain alloys are separated based on boiling points – an example being silver that is extracted from its ores when alloyed with zinc. This is called the Parkes process, and is shown in Figure 4.11. It is not a stoichiometric process, rather one that depends on both miscibility and boiling points. In the process, liquid zinc is mixed with liquid elemental lead. These two metals in their molten state are essentially immiscible, but silver metal, impurities of which are in the lead, are highly miscible with molten zinc. The end result is that silver amalgamates with the liquid zinc, resulting in pure lead (the main product). The resulting zinc–silver alloy is then placed in a retort and heated until the zinc vaporizes. The valuable silver is thus recovered, and the zinc is cooled and recaptured for reuse. This final separation occurs because the boiling point of zinc is 907 °C, while that of silver is a much higher 2,162 °C.

$$Pb{:}Ag_{(l)(alloyed)} + Zn_{(l)} \rightarrow Zn{:}Ag_{(l)} + Pb_{(l)(pure)}$$

$$Zn{:}Ag_{(l)} \rightarrow \Delta \rightarrow Zn_{(g)} + Ag_{(l)}$$

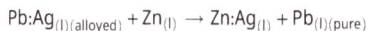

Figure 4.11: The Parkes process.

4.3 Secondary bonding, intermolecular forces

Beyond covalent, ionic, or metallic bonding, there are several interactions between atoms that have an influence on a material, and on that material's characteristics and macroscopic physical properties. We discuss in this section those intermolecular forces that can have pronounced influence.

What are broadly called van der Waals forces were first studied by Johannes Diderik van der Waals, for whom they were named. These are the intermolecular forces beyond covalent or ionic bonding, and can be considered weak when looked at as an individual attraction between two atoms or two positions in a molecule or molecules. But when summed enormous number of times, their strength can be significant. This becomes important when dealing with molecules such as synthetic polymers, or with proteins.

All secondary, intermolecular forces arise because of dipoles within molecules, ions, and atomic units, whether they are permanent or temporary. Ionic and covalent forces are always strong enough that their attractions dominate over the intermolecular forces. Still, these intermolecular forces – hydrogen bonding, dipole attractions, and London dispersion forces – do manifest themselves in some macroscopic, physical properties, which we discuss here.

4.3.1 Hydrogen bonding

Hydrogen bonding, the bond between a hydrogen atom on one molecule and an oxygen, nitrogen, or fluorine atom on another is the strongest form of intermolecular attraction [6]. The hydrogen bond occurs because the covalent bonds between a nitrogen atom and a hydrogen atom, or an oxygen atom and a hydrogen atom, or a fluorine atom and a hydrogen atom are polar enough that the hydrogen atom is partially positive in the direction away from the bond to that just mentioned larger atom. In addition, these three atoms, N, O, and F, have sufficient electronegativity and electron density about them that they can be said to present a partially negative end or side to other hydrogen atoms in a sample of the molecules. Figure 4.12 shows examples of this.

While simple examples of hydrogen bonding, such as those shown above, are common and cited in numerous chemistry texts, there are many examples of what might be termed dual-molecular-type hydrogen bonding as well. For example, by mix-

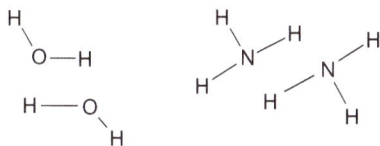

Figure 4.12: Hydrogen bonding in water.

ing water and ethanol, which have O–H covalent bonds and hydrogen bond, results in total volume of mixing less than the sum volumes of the two starting liquids. What we mean, for example, is that a combination of 5.0 mL of water and 5.0 mL of ethanol does not mix and results in 10.0 mL of solution. The total volume is approximately 9.3 mL of the mixed liquid.

Some molecules can undergo hydrogen bonding at multiple positions within the same molecule. One common material that can do so is ethylenediaminetetraacetic acid, usually abbreviated EDTA (Figure 4.13). The four carboxylic acid moieties are all capable of hydrogen bonding, although this molecule is routinely used as a chelating agent. This means that some cation in solution will be attached to one or more of the carboxylic acids in EDTA, resulting in a coordinate covalent bond, as we have discussed earlier.

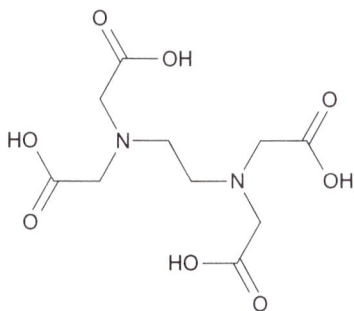

Figure 4.13: Lewis structure of ethylenediamine tetraacetic acid (EDTA).

There are also numerous examples in which a molecule has one or more intramolecular hydrogen bonds. Figure 4.14 shows this in the molecule generally known as salen, which we have discussed earlier and shown in Figure 4.9. Here in Figure 4.14 we have emphasized the intramolecular hydrogen bonds between the –OH and the imine nitrogen atoms in the molecule.

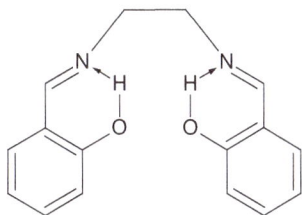

Figure 4.14: Salen with intramolecular hydrogen bonds.

The formation of this molecule, as shown in Figure 4.14, is driven in part by the formation of these two stable intramolecular hydrogen bonds, as well as from the Schiff's base condensation between a primary amine (ethylenediamine) and an aldehyde (salicaldehyde). The attraction is strong enough that the molecule will form from the mixing of the two reactants as neat liquids at ambient temperature. In virtually all syntheses of this, a solvent is used so that formation of the product, which is a solid, does not hinder the reaction proceeding to completion.

One of the largest molecules in which multiple hydrogen bonds are part of the molecule is DNA.

The alignment of adenine to thymine, and of cytosine to guanine, is referred to as the alignment of complementary base pairs and is shown in Figure 4.15. The A–D alignment is made possible by two hydrogen bonds. The C–G alignment has three hydrogen bonds in it. These sets of hydrogen bonding, repeated thousands of time, are what holds DNA together, and what forms the twist of the double helix, even though a single hydrogen bond is much weaker than a single covalent bond.

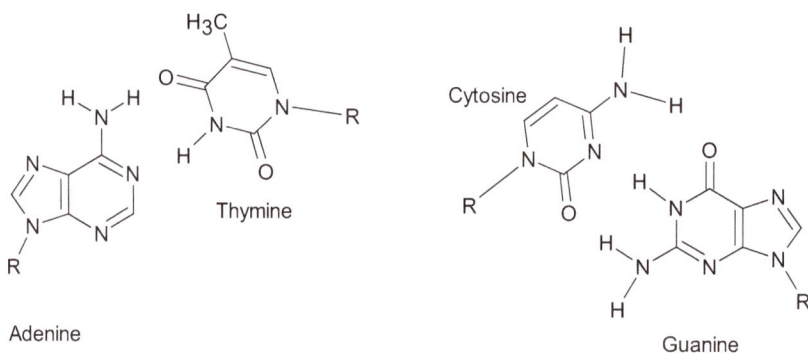

Figure 4.15: Complementary base pairing in DNA.

4.3.2 Dipole interactions

Polar molecules by definition have at least one positive end or point or side, and one or more negative points – and these points have an effect on other molecules or ions that are closest to them. An example is the common organic solvent "methylene chloride," aka. dichloromethane, shown in Figure 4.16. The molecule is a distorted tetrahedron in shape with a negative edge between the two chlorine atoms, and a positive edge between the two hydrogen atoms. Perhaps obviously, one molecule of it will align positive to negative with the molecule immediately next to it. In Figure 4.16, an arrow has been included. The arrowhead is pointed in the direction of the partially negative side of the molecule, while the tail is at the partially positive side. Another notation for this is to use δ– and δ+ (said as "delta-negative" and "delta-positive"), where the head and tail of the arrow would be.

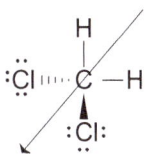

Figure 4.16: Methylene chloride with arrow pointing to negative edge.

Dipolar attractions also exist between two polar but different molecules. An example of two common materials for which this attraction occurs is acetone and water. Figure 4.17 shows one possible way in which the two molecules can align so that there is maximum attraction from the positive portion of one molecule to the negative portion of the other.

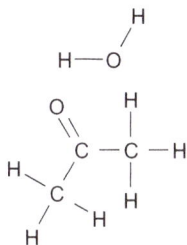

Figure 4.17: Acetone and water intermolecular alignment.

These intermolecular attractions are not as strong as the just mentioned hydrogen bonding, but are still measurable. Much like the volume contraction mentioned in Section 4.3.1, mixing various amounts of acetone and water will result in total volumes that are less than the sum of the two-component liquids. This is not a matter of hydrogen bonding, but the end result is a similar decrease in volume.

   Dipoles can also be induced in a material. Perhaps the simplest or most common example is common table salt dissolved in water. The sodium ions align water molecules around them with the electron-dense oxygen atom becoming closest to the cation. In like fashion, the chloride anion aligns water molecules around it, this time with the partially positive hydrogen atoms aligned closest to the anion. Figure 4.18 shows this in two dimensions, although perhaps obviously, it occurs in three.

4.3.3 London forces

Generally considered the weakest of the intermolecular forces, London forces were first studied by Fritz Wolfgang London, who began his work on the subject in the 1930s. They are best described as an interaction between either atoms or molecules

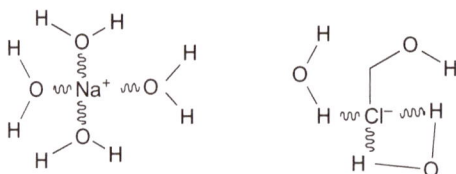

Figure 4.18: Induced dipoles in water.

that is dependent on the distance between atoms and that affects even molecules that are not polar. Such attractions are caused by variations in electron density within molecules, and then between molecules. Such transient dipoles can be induced by specific environmental conditions in the material.

As with the discussion of molecular size, in Section 4.1.2, or the discussion of hydrogen bonding in Section 4.3.1, London forces can be thought of as having an additive effect when there are numerous sites for them – effects which manifest themselves in physical properties such as melting point or boiling point. For example, there are very few sites for intermolecular bonding in the very lightweight, nonpolar hydrocarbon methane, which is a gas at room temperature. The hydrocarbon "pentane" is a liquid at room temperature and is a larger molecule, with more sites for such forces to act. When hydrocarbon molecules contain 30 or more carbon atoms – when they are considered paraffins – they are waxy solids at room temperature. Thus, melting point and boiling point increases with the increasing molecular weight of hydrocarbons, and the corresponding increase in the number of sites at which London forces can occur. Figure 4.19 shows this schematically.

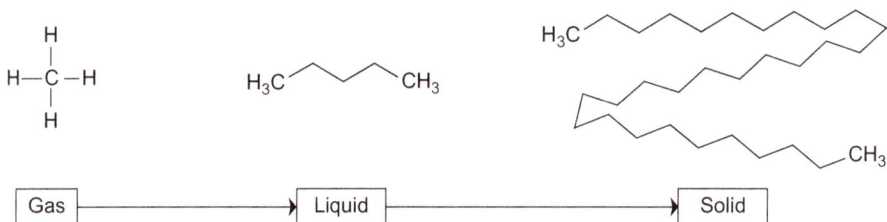

Figure 4.19: Trend in melting points and molecular weights of hydrocarbons.

van der Waals forces and London forces exist in other molecules besides hydrocarbons. But in organic molecules that also have dipoles, such as acetone and water, which have been mentioned above, the strength of their dipoles or of their hydrogen bonding is significantly greater than the London forces.

References

[1] R.C. Maurya. Inorganic Chemistry: Some New Facets, 2021, Walter DeGruyter, GmbH. ISBN: 978-311072725-8.

[2] M.A. Benvenuto. Industrial Inorganic Chemistry, 2015, Walter DeGruyter, GmbH. ISBN: 978-311033032-8.

[3] P. Kurz, N. Stock. Synthetische Anorganische Chemie, 2013, Walter DeGruyter, GmbH. ISBN: 978-311025874-5.

[4] M.A. Benvenuto. Metals and Alloys, Industrial Applications, 2016, Walter DeGruyter, GmbH. ISBN: 978-311040784-6.

[5] R. Steudel. Chemistry of the Non-Metals, 2020, Walter DeGruyter, GmbH. ISBN: 978-311057805-8.

[6] A. Huettermann. The Hydrogen Bond: A Bond for Life, 2019, Walter DeGruyter, GmbH. ISBN: 978-311062794-7.

Chapter 5
Structure of matter

5.1 Crystalline solids

A great number of crystalline solids exist in nature, with all of their structures falling into a relatively small number of stacking and packing arrangements. The reason the number of packing arrangements is relatively small compared to the number of crystalline materials that occur naturally is because crystals are often composed of one or two atoms or ions in a neatly repeating fashion, and there are only a limited number of ways by which two different sized spheres can pack with each other – or in the case of pure elements, that spheres of the same size can pack together. This being said, more complex molecules can stack and form crystalline solids as well, and elemental solids can form crystals also. Figure 5.1 shows a crystalline amethyst mineral sample as an example of one material that crystallizes beautifully in nature.

Figure 5.1: Amethyst crystals.

Dividing all crystalline materials into two broad categories, we arrive at ionic materials and metallic materials. There are some nonmetallic elements that do crystallize as well, but the substances most chemists and engineers will encounter and be most familiar with fall into the two broad categories. It should be noted though that ionic materials which crystalize are not only diatomic materials, such as NaCl. Much more complex salts can do so as well, such as barium carbonate, or calcium titanate, or ammonium phosphate, to name three examples – one of which includes covalent bonds. Similarly, while elemental metals routinely can be crystalized, so can many metal alloys [1].

https://doi.org/10.1515/9783112205822-005

5.2 Unit cells and crystals

Crystalline materials have repeat patterns which are called unit cells. The unit cell of a crystal is the smallest group of atoms or particles that contain the overall symmetry of the material, and that when repeated will become the overall crystal pattern. From the unit cell a complete lattice can be constructed by constant repetition of the unit cell into three dimensions.

Table 5.1 shows the crystal systems into which all crystalline solids can be included. The most common type, that found in most crystalline materials, is the monoclinic system. Nearly 1,500 crystals have been identified as being in this system.

Table 5.1: Crystal systems.

Name	Definition of unit cell definition	Dimensional equalities	Relations among lattice parameters	Example figure
Cubic	A	All sides equal, all angles 90°	$a=b=c$ $\alpha=\beta=\gamma=90°$	
Tetragonal	a, c	Two sides equal, all angles 90°	$a=b\neq c$ $\alpha=\beta=\gamma=90°$	
Hexagonal	a, c	Two sides equal, 2 angles 90°, 1 angle is 120°	$a=b\neq c$ $\alpha=\beta=90°\ \gamma=120°$	
Rhombohedral, aka trigonal		Three sides equal, 2 angles 90°	$a=b=c$ $\alpha=\beta=\gamma=90°$	
Orthorhombic	a, b, c	No sides equal, 3 angles 90°	$a\neq b\neq c$ $\alpha=\beta=\gamma=90°$	
Monoclinic	a, b, c, β	No sides equal, 2 angles 90°	$a\neq b\neq c$ $\alpha=\gamma=90°\ \beta\neq90°$	
Triclinic	a, b, c and α, β, γ	No sides equal, no angles equal	$a\neq b\neq c$ $\alpha\neq\beta\neq\gamma\neq90°$	

Note that while symmetry decreases from the most symmetrical of the seven systems, the cubic structure, in which all sides are the same length – the a, b, and c – and all bond angles in all three directions are the same – the α, β, and γ – there are still elements of symmetry in all seven of the systems.

These seven basic crystal systems have examples of the unit cell shapes shown in the right-hand column of Table 5.1. The notation normally used for crystal types is *a*, *b*, and *c* for the sides of a crystal, and the small Greek letters – alpha, beta, and gamma – for the angles at each corner, often written as *α, β, γ*. Note the instances in which lengths of sides are the same, or angles are the same. For example, in the cubic system, since all sides are of the same length, as are all the angles, only one dimension of distance needs be known to determine the size of the crystal.

Additionally, in several of the cases in Table 5.1, the angle at some corner of the unit cell is 90°. This does not automatically make that cell cubic – although perhaps obviously any cubic structure must have all 90° angles.

5.3 Packing of particles

Since atoms can be considered spheres, at least in practical terms and in the idea of them packing together as solids, a method of determining the structures of solids can be determined based simply on that packing. This method is applicable to all crystals in which all the particles are the same. Thus, this is useful for describing the packing and crystal structure of all elemental, crystalline solids.

It is sometimes helpful to compare, either in one's mind or by direct observation, the way fruits are packed and displayed in the produce section of a grocery store. They are generally stacked in only two ways. In one, a single piece of fruit is surrounded by and touching only six others, if they are all laying on the same surface. In the other, a single piece of fruit is touching only four others if all are on the same surface (of course, there are also bins in which fruit is randomly placed).

In similar vein, one can observe the way cannon balls are stacked and displayed at national parks and monuments. Both of these examples show how particles of the same size are packed with other particles of the same size with minimal empty or wasted space. Figure 5.2 shows an example of this macroscopic version of spherical packing.

A fairly straightforward way to examine crystalline structure is thus through the packing of spheres in layers. Figure 5.3 shows a schematic of a single layer of spheres – which we can easily envision to be atoms – lying on some surface. Note that in Figure 5.3 we have shown more than the minimal number of spheres.

The arrangement in Figure 5.3 is as close together as an array of spheres can get to one another (we will see a second arrangement shortly, in this discussion) – six spheres touching one sphere in all cases. In between the atoms or spheres we can see

Figure 5.2: Packing of produce in a grocery store.

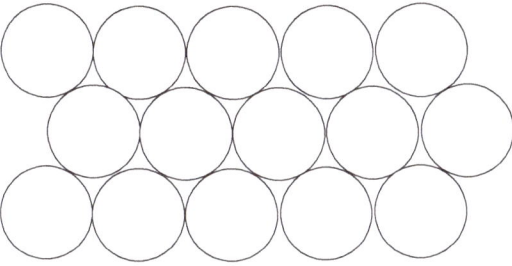

Figure 5.3: Layer of spheres.

triangular spaces or voids. A second layer of spheres can be placed atop the first, as shown in Figure 5.4.

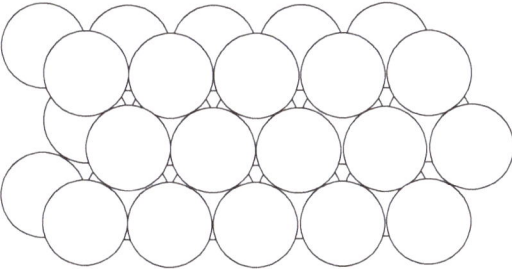

Figure 5.4: Two layers of spheres.

Note that the second layer only covers one half of the voids between the spheres of the first layer. In doing so, the covered spaces now form a three-dimensional void that is the shape of a tetrahedron. The other half of those void spaces remain open and can be "seen" from the top to the bottom in Figure 5.4. After this second layer is placed, a third layer can be added, either directly over the first layer, or shifted so that the remaining half of the open spaces are covered. If the remaining voids are covered, the three-dimensional shape of that void is now an octahedron. The first arrangement is often called an A–B layer, or sometimes an A–B–A layer arrangement, since the third and first layers are positioned the same. The second is usually termed an A–B–C arrangement.

The A–B layer arrangement is called hexagonal close packing, or hcp.

The A–B–C layer arrangement is called cubic close packing, or ccp. This corresponds to a face-centered cubic arrangement, which we discuss below.

Another means by which crystalline structures are characterized is by the atoms in each unit cell, of a cube shape. Figure 5.5 shows the simplest form of this, what is called simple cubic, or in some texts, primitive cubic. Each corner of the cube is where an atom is positioned. One can note that 1/8th of each corner atom is in the cube, and therefore a total of one atom makes up the unit cell. It is actually rather rare in nature because there is a significant amount of empty space in the lattice.

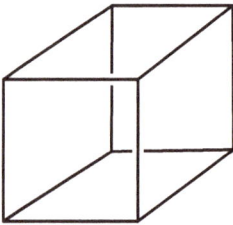

Figure 5.5: Simple cubic structure.

Figure 5.6 shows what is called the face-centered cubic structure (fcc), and the body-centered cubic structure (bcc). The fcc structure can be likened to a gambling die, in which a dot or dots are on each face, and with atoms still in the corner positions. Thus, there is still 1/8th of an atom in the cell from each corner, but six face atoms that are ½ in the cell. There are therefore four atoms total in the cell. The bcc can be likened to a ball in a box, with the ball representing one atom completely in the cube, and eight atoms again positioned at the corners. Since the atom in the center is completely in the cell, there are a total of two atoms in the unit cell.

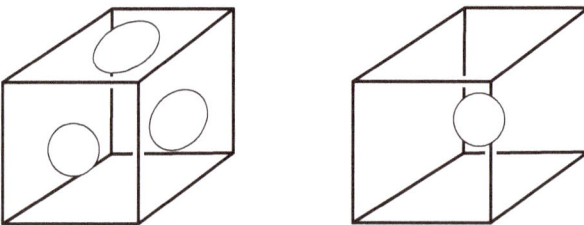

Figure 5.6: Face-centered cubic structure and body-centered cubic structure.

Determining the radius of an atom in such a unit cell is a simple matter of measuring a side and dividing by two for a simple cubic structure – and for determining the side of a unit cell a matter of using basic trigonometry for a fcc or bcc structure. In the fcc structure, since the distance between two corners is 4r – because the atoms in the corner and on the face touch each other – the formula to find a radius is simply dividing the hypotenuse by 4. To determine the side of the unit cell, which is one side of an isosceles triangle, the ratio of side-side-hypotenuse is $1 \times 1 \times \sqrt{2}$.

For a bcc arrangement, the distance between opposing corners is $\sqrt{3} \times$(side). Assuming all the atoms are the same, again the radius is ¼th of that distance.

Beyond the examples mentioned above and their atomic arrangements, characteristics of crystalline packing can be seen in macroscopic materials, and can often be found in minerals. Crystalline calcite, for example, is $CaCO_3$ in a rhombohedral structure that can be seen in macroscopic examples of the mineral. Figure 5.7 shows a photograph of this.

Figure 5.7: Natural calcite crystal.

5.4 Holes in crystal structures

We will discuss the concept of holes, sometimes called defects, later in the book, in Chapter 9, but at this point will state that irregularities in crystal packing can be caused by missing particles from regular positions – holes – or particles added in to the spaces that cannot then be occupied by the particle spheres that make up the material [2].

The presence of some type of defect in any crystal does not automatically change the properties of the material completely. But even small amounts of some added element in the material's matrix – or the removal thereof – can change one or more properties.

Using a famous diamond as an example, the Hope diamond has a blue tint to it. The diamond matrix, a C_{10} repeating unit, produces no color, and thus the purest of diamonds are clear. The blue is imparted by the presence of a small amount of boron in the diamond. The diamond is still hard (and very valuable), but now has a different color.

5.5 Amorphous materials

Solid materials that can exist in some crystalline form always can exist in some amorphous form as well. Additionally, what is called poly-typism occurs in some elements, as well as in some compounds.

Elemental sulfur is an excellent example of a substance having both a crystalline form as well as an amorphous one. What is called crown sulfur, S_8, exists in an ordered array shaped like a crown, for which it is named. Yet "plastic sulfur," as amorphous sulfur is sometimes called, exists as a solid with no real order. A classic chemical demonstration that has now been largely discarded among professors is to melt powdered sulfur in a crucible until it glows a dull red, then quench it by pouring it directly into a cold water bath. The resulting material looks somewhat like spaghetti, and is very stretchable. Hence the name, plastic sulfur. Figure 5.8 shows an example of this.

Figure 5.8: Plastic sulfur.

Beyond elements, materials such as glass, rubber, and plastics are often amorphous. These three examples indicate the wide array of different physical properties that amorphous materials can have. Glass, for example, only flows at high temperature and is rigid when broken. Rubber is often a runny, viscous material, especially when not vulcanized. And plastics span a range from highly flexible to extremely rigid – examples of the first being atactic polypropylene and of the second being isotactic polypropylene.

5.6 Semisolids

The term "semisolid," or "quasi-solid," is one that indicates an obvious and dominant property of the material which it describes. All such materials have in common a viscosity that is high enough that the material is not considered a liquid, and does not flow like one, but low enough that it is not often considered a solid. Examples of common semisolids include the following:

1. Butter and margarines
2. Peanut butter
3. Creams for personal care products
4. Gels – edible gelatins
5. Greases – nonedible, for machine lubrication, as well as edible
6. Mayonnaise
7. Waxes

This list is not all inclusive.

Often, semisolids are made and first used in some state of elevated temperature, so that the material may flow into a desired shape or container. A perhaps obvious example is peanut butter, which is put into its containers when it is hot enough to flow, but which does not flow when a consumer purchases and uses it. But many other semisolids that are not food products are treated the same way, such as gel packs that are used as a base layer under certain types of protective body armor.

One property that semisolids share, one that makes them different on a molecular level from crystalline materials, is that they have no well-ordered repeat structure. Many times this is because their molecular structure is of larger, long molecules that do not align with each other in any way. Numerous fats and fatty acids are components of semisolids. Unsaturated fats – those having one or more double bonds – tend to be more liquid at room temperature, while saturated fats – those with only single bonds in their chains – tend to be more solid.

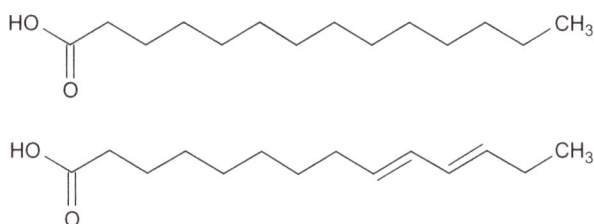

Figure 5.9: Examples of Lewis structure of a saturated fat, and a polyunsaturated fat.

5.7 Crystalline or noncrystalline metallic solids

Beyond what we have discussed in Section 5.3, we will defer much of the discussion of metallic solids to Chapter 8, but will say here that metallic elements too can form either amorphous solids, or crystalline materials. Often, the form a metal takes depends upon how it is solidified from the melt [3]. We have mentioned several examples, above.

Magnetic iron probably remains the most obvious and common example of how a material behaves when it possesses some crystalline structure as opposed to none. When a sample of iron has domains or areas in which electron spins are aligned, it is magnetic – it exhibits what is known as ferromagnetism. If a magnetic sample like this is heated strongly, then quenched, simply by immersion in ice water, the magnetism is lost.

5.7.1 Mercury and gallium

Two metals that have melting points near or under what is generally called room temperature are elemental mercury and elemental gallium. Elemental mercury has a melting point of −38.8 °C. It is well known that this is a liquid at room temperature. Indeed, its ancient name is related to it being a liquid at room temperature: quicksilver.

Gallium has a much shorter history, having only been isolated and discovered in 1875, and having a melting point of 29.7 °C. It can often melt when a person holds it in their hand. A niche use for gallium is the production of the low-melting alloy galinstan, a mixture of gallium, indium, and tin (hence the name). The composition is 68.5% gallium, 21.5% indium, and 10.0% tin, and the alloy melts at −19 °C. Because of this low melting point, the alloy has been found to be useful in specialized, low temperature thermometers.

5.7.2 Tin scream

While several elemental metals exhibit poly-typism, meaning they have two or more allotropes, tin is one of only a few metals that has two, and that exhibits an audible phenomenon when subjected to stress. Relatively thin bars of tin can be bent, and when this happens a crackling sound can be heard. Termed, "tin scream," or "tin cry," this occurs because crystals in the tin are being disrupted from their alignment. The phenomenon can be repeated several times with a single bar of tin. When the cracking is no longer heard as the bar is being bent, the bar can be melted, re-cast, allowed to cool slowly, and the crystals re-formed so that tin scream can again be heard in the sample.

This type of "cry" can also occur with indium, zinc, and gallium.

5.7.3 Metal foams

A relatively new field of study within materials chemistry is the production of metal foams. A metal foam is a material much like a metal sponge, in which the metal atoms are positioned akin to the solid sponge material, and air occupies the leftover space. The metal framework is rigid however, unlike a normal sponge. Materials have been created in this way in which the overall material density is extremely low since the framework metal atoms are amidst a significant volume of air, at times over 75%. In the field, metal foams are divided into two classes, closed cell and open cell. A large number of them are produced using aluminum, since it is an extremely low density metal.

To produce metal foams, some gas or some foaming agent is introduced to a molten metal which then produces the foam during the cooling phase. During such production, caution must be taken to ensure the bubbles remain in the cooling metal material long enough to ensure foam formation. When the metal exists as a viscous material in the melt, injecting gas into that material is a useful and productive technique.

Uses of metal foams include heat exchangers as well as shock absorbent materials. Unfortunately, production of metal foams continues to be expensive, despite an established history for such materials (initial patents were issued as far back as the 1940s). Because of the expense of their manufacture, foams tend to occupy niche applications, and not large-scale ones. As an example, lightweight materials are required by the aerospace industry in numerous applications. Metal foams are often applicable here [4–6].

5.8 Recycling

Materials have not in the past been recycled because they are in some way crystalline. Several types of metal and glass have been recycled at various times, but this is because they are useful commodities in many forms, and because their recycling and reuse is less expensive than the production of new, virgin materials. It has not been because of their crystallinity.

The recycling of metals is an old and established practice that predates any environmental movement. America's first recycling drive for metals was during the First World War, and was aimed at raising the awareness of Americans for materials needs for what was at the time considered to be a European war. The second major recycling effort came during the Second World War. This effort was indeed aimed at obtaining more metal for the war effort.

Much more recently, it can be argued that recycling efforts for metals that consumers use are based on the fact that much more aluminum is now used to make beverage cans than in the past. Originally, soda pop, beers, and other beverages were sold in glass bottles, often with corks as the top. Over time, this evolved into aluminum can for beverages. Refining aluminum from bauxite ore is always an energy-intensive process – the Hall–Heroult process – and recycling aluminum is far less expensive than continually refining more.

Recycling various materials has now become normal in some communities and in some businesses. Not only are materials recycled, whether they are crystalline or not, but user end items are recycled, often because of the value of the material that makes up such items. Figure 5.10 shows a typical recycling receptacle that receives multiple materials and items.

Figure 5.10: Recycling bin in an electronics store.

The recycling of metals such as steel, copper, brasses, and bronzes is also an established industry, but is not seen by consumers as often as the recycling of aluminum is. Scrap yards often deal with large amounts of these metals, buying from other businesses, then either selling directly, or performing some melting, purification, and reforming before selling.

References

[1]   S. Schorr, C. Weidenthaler. Crystallography in Materials Science, 2021, Walter DeGruyter, GmbH. ISBN: 978-311067485-9.

[2]   W. Neumann, A. Mogilatenko, K. Scheerschmidt. The Nature of Crystal Defects, 2022, Walter DeGruyter, GmbH. ISBN: 978-311062150-1.

[3]   F. Hensel, W.W. Warren. Fluid Metals, the Liquid-Vapor Transition of Metals, 2014, Walter DeGruyter, GmbH. ISBN: 978-1400865000-0.

[4]   N. Dukhan. Metal Foams: Fundamentals and Applications, Lancaster, PA, USA, DEStech Publications. 978-1-60595-014-3.

[5]   K. Stoebener, J. Baumeister, G. Rausch, M. Busse. Metal foams with Adbvanced Pore Morphology (APM), *High Temperature Materials and Processes*, 2007, 26(4): 231–238.

[6]   G. Lange. Metallschaeume: Herstellung, Eigenschaften, Potenziale Und Forchungsansaetze – Mit Schwerpunkt Auf Aluminumschaeume, 2020, Walter DeGruyter, GmbH. ISBN: 978-311068155-0.

Chapter 6
Molecules

6.1 Introduction

Distinct molecules – combinations of atoms that exist in precise, stoichiometric ratios – can exist in the gas, liquid, and solid phase, always being composed of two atoms or more in an electrically neutral configuration held in their positions routinely by some covalent bond, although ionic, electrostatic, and dative bonds can be part of them. Material chemistry tends to deal with molecules in the solid state, although techniques such as chemical vapor deposition (CVD) and selective crystallization involve the transfer of molecules or atoms from the gas phases or from a solvated state to that of a solid, and sublimation describes the reverse physical change from solid to gas.

6.2 Organic molecular materials

Organic molecules exist in all three phases of matter, but in this book we will focus much of our examination on those which exist as solids at ambient temperature and pressure, with some notable exceptions. The study of organic matter in the solid phase bears certain similarities to inorganic matter in the same phase, such as how molecules pack. But there are significant differences as well. For example, the ability of carbon to polymerize into chains thousands of carbon atoms in length appears to be both unique to the element, and extremely well studied, as several polymers are manufactured in enormous quantities each year – and will be discussed in more detail in Chapter 7.

6.2.1 Fully covalent molecules

The view of single, discrete molecules presented to students when they begin a college-level study of chemistry is that they are generally small units that are held together internally by covalent bonds, and from one molecule to another through some intermolecular force or forces, as discussed in Chapter 4. Examples of purely covalent molecules, in which valence electron density is evenly shared, are usually confined to oxygen, nitrogen, hydrogen, and the column of halogens.

While the above is true as a definition of covalent molecules, it is probably fair to say it is not a complete description of molecular materials.

Using a series of progressively heavier hydrocarbons as an example, all of which are covalently bonded molecules, the longer a linear hydrocarbon becomes, the more

https://doi.org/10.1515/9783112205822-006

sites exist for it to have enhanced intermolecular forces, such as London forces. This manifests itself in its physical form, going from gases at the lightest molecular weights, to solids for the heaviest. Examples in a nonexhaustive list are shown in Table 6.1.

Table 6.1: Hydrocarbon molecular weights and physical states.

Name	Formula	m.p. (°C)	Physical state
Methane	CH_4	−182.5	Gas
Ethane	C_2H_6	−182.8	Gas
Propane	C_3H_8	−187.7	Gas
Butane	C_4H_{10}	−140.0	Gas
Pentane	C_5H_{12}	−129.8	Liquid
Hexane	C_6H_{14}	−96.0	Liquid
Octane	C_8H_{18}	−70.6	Liquid
Decane	$C_{10}H_{22}$	−30.5	Liquid
Eicosane	$C_{20}H_{42}$	36–38	Solid

The uses of these materials is wide, and is generally dependent upon whether they are gases, liquids, or solids – and is not always dependent upon whether a hydrocarbon is pure, or a mixture of isomers, or a mixture of molecules of roughly the same molecular weight. The general public is arguably most familiar with liquid hydrocarbons that find use as motor fuels.

In addition to lower-weight hydrocarbons, there are numerous polymers that have been made in the past seventy years, all with extremely high molecular weights by comparison. But only six of these polymers are so ubiquitous that they each have their own resin identification code, or RIC. Polystyrene is RIC 6, and is often thought of as Styrofoam® by the general public. Figure 6.1 shows the Lewis structure of the repeat unit of polystyrene, and is shown here as an example of a molecule that can have an extremely high molecular weight, but still be composed entirely of covalent bonds, thus making it a single molecule.

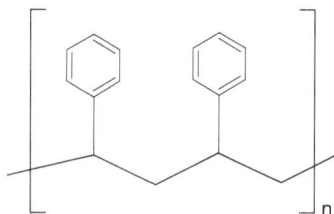

Figure 6.1: Lewis structure of the repeat unit of polystyrene.

Note, there is a fourth bond at each carbon atom that is already connected to a phenyl group in Figure 6.1, but we have not included it in the figure, and therefore not shown any directionality to the phenyl groups – meaning, not indicated if they are

angled in front of or behind the main carbon chain. This directionality – called tacticity – is important for the properties of this and other materials, but does not affect
whether or not the polystyrene is a molecular material or not.

The covalent bonds which exist in smaller hydrocarbon molecules (and indeed, in all
small, covalently bonded molecules) are no different from those in polymers. Figure 6.2
shows the two isomers possible for C_4H_{10} – n-butane and isobutane – both of which are
gases at ambient temperature and pressure.

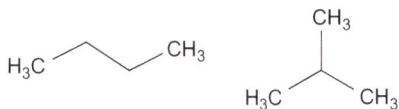

Figure 6.2: Lewis structures of n-butane and isobutane.

Heavier hydrocarbons can exist as solids, even though they are not as high in molecular weights as polymers. As an example, Figure 6.3 shows the Lewis structure of myricyl palmitate, one of the major components of beeswax. Note that it does contain an
ester functionality, but is largely a linear hydrocarbon molecule.

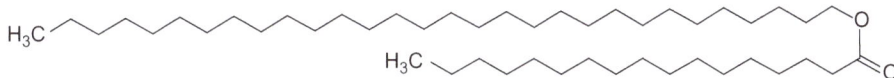

Figure 6.3: Lewis structure of myricyl palmitate, a wax.

6.2.2 The organic–inorganic matter borderline

The definition of organic chemistry has often been: "the chemistry of carbon," although very few chemists or engineers would argue that the chemistry of diamond or
graphite would qualify as part of organic chemistry. Likewise, the chemistry based on
the carbonate anion (CO_3^{2-}) would generally not be included in such a definition.

We point out this apparent border of the field of organic chemistry to reinforce
the idea that materials chemistry includes substances that might have traditionally
been included in more than one subfield.

A somewhat different definition of organic chemistry might be that of "the chemistry of carbon and the carbon–hydrogen bond." This certainly falls well within the
scope of molecules as we discuss them in this chapter. But this definition also has
fuzzy edges. For example, does carbon tetrachloride, CCl_4, then qualify as being organic, since there are no carbon–hydrogen bonds in it? Does perfluoroethylene (C_2F_4)
qualify as organic or inorganic? The production of this small molecule, shown in Figure 6.4, has become a large industry and a large part of materials chemistry, in the it
is the precursor to Teflon® and the many materials made from it.

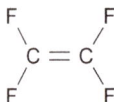

Figure 6.4: Lewis structure of C_2F_4.

Other examples of this area that can be considered organic chemistry, but that might be considered inorganic chemistry are the chlorofluorocarbons, or CFCs. Two examples are shown in Figure 6.5: one methane-based, the other ethane-based. But essentially all possible permutations and isomers of the one-carbon and two-carbon CFCs have been prepared.

Figure 6.5: Examples of CFCs.

Uses of CFCs have been as foams for a wide variety of packing materials, and as refrigerants. They have been made in industrial-scale quantities for such industries, but in the past decades have been phased out in many countries because of the harm they do to the planetary ozone layer.

A further field that straddles the organic–inorganic border is the silicones. These are covalently bonded polymers, much like the polymers with carbon chain backbones, like polyethylene or polypropylene, but their main chains are alternation silicon and oxygen atoms. Strictly by that character, they are inorganic polymers. But the material properties of silicones on a macroscopic basis are greatly affected by the side chain moieties pendant to each silicon atom. Two examples are shown in Figure 6.6. The short methyl groups result in a silicone with very different properties than that with the much longer *n*-hexyl groups. This argues that such molecules are inherently organic in their composition.

6.2.3 Waxes

We have seen the Lewis structure of a major component of beeswax, above, but not all waxes have to be linear chains of ester-containing hydrocarbons. A wide variety of hydrocarbons can be considered waxes, some simply as long alkanes of varying chain length, including those shown in Figure 6.7. As well, other organic molecules can be considered as waxes, including that shown in Figure 6.8, the molecules generally known as petrolatum.

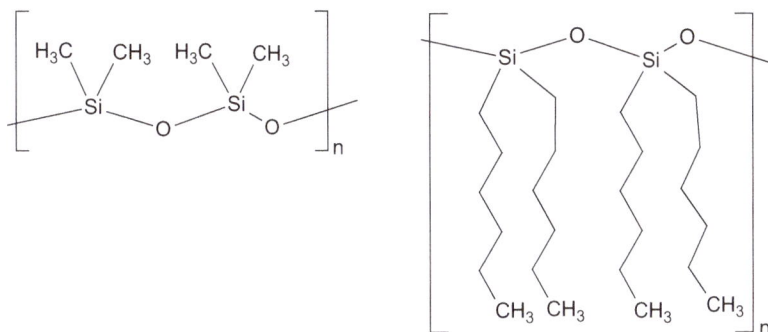

Figure 6.6: Examples of the Lewis structures of silicones.

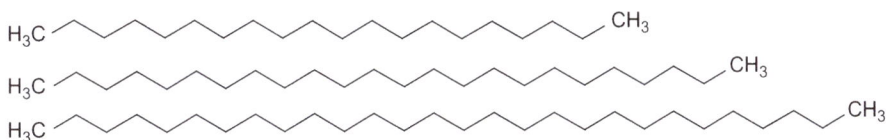

Figure 6.7: Lewis structure of hydrocarbon waxes.

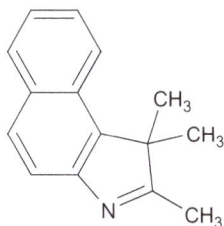

Figure 6.8: Lewis structure of petrolatum.

We can see that waxes have different sources. Beeswax is rather obviously a biological product, and thus a biological source. Alkane waxes and petrolatum are sourced from petroleum. The common factor these molecular materials have is that they are all solids or semi-solids with a waxy appearance and feel.

6.2.4 Polymers

Both natural and synthetic polymers expand the definition of molecule somewhat, because the original, historical definitions did not account for repeat units of thousands or tens of thousands of a basic moiety, the result being a single molecule. As an example, Figure 6.9 shows the repeat unit of polyethylene and polypropylene, emphasizing the difference in the methyl units that are pendant to the main chain of the polymer for polypropylene, the lower part of the figure. Figure 6.10 shows the repeat unit of natural rubber. This is a polymer that has been known for centuries, but its molecular structure was not determined for most of that time.

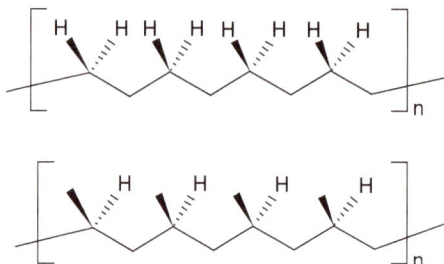

Figure 6.9: Repeat units of polyethylene and polypropylene.

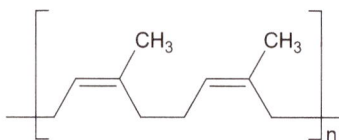

Figure 6.10: Repeat unit of rubber.

6.3 Carbon dioxide

This one molecule stands at something of a crossroads or nexus of organic and inorganic chemistry, as it is often the end product of some organic combustion, as shown in Figure 6.11, but can also be generated in ways that utilize a source which can be classified as inorganic.

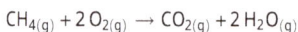

$$CH_{4(g)} + 2\,O_{2(g)} \rightarrow CO_{2(g)} + 2\,H_2O_{(g)}$$

Figure 6.11: Example production of CO_2.

The gas that is generated can be captured, refrigerated and pressurized (since CO_2 can liquefy at <6 atm), resulting in liquid carbon dioxide. When a quick pressure reduction follows, this vaporizes some portion of the material – and in the process cools the remainder enough that it solidifies – becomes dry ice. This resulting dry ice can then be formed into pellet-sized pieces or larger blocks.

By whatever means CO_2 is generated though, the amount of it that has been released into the atmosphere since the Industrial Revolution has altered the atmospheric chemistry of the planet. Its large-scale removal, by any of a variety of means, would be advantageous. Figure 6.12 shows the basic chemistry of how carbon dioxide can be immobilized as oxalate anion. This has been studied, and found to occur in some natural systems, such as species of cacti. But it has not yet been brought up to an industrial scale capable of cancelling out the amount of CO_2 generated industrially each year [1, 2].

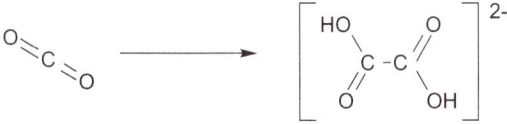

Figure 6.12: Production of oxalate from carbon dioxide.

As a stand-alone molecule, CO_2 does find numerous uses. Carbonated beverages may seem like a small application, but the production and sale of soft drinks is an enormous industry that always utilizes carbon dioxide on an industrial scale. Figure 6.13 illustrates an example of this, at least in carbon dioxide is moved. A more complete list of its uses includes the following major applications:

– Food cooling, storage, and long-term preservation
– The just-mentioned carbonating of beverages.
– Blast cleaning
– Enhanced oil recovery via CO_2 injection into wells forcing oil from geologic formations.
– Blanketing or inert atmosphere for storage of grains
– Adjusting pH of wastewater streams

Figure 6.13: Carbon dioxide transported by truck.

Beyond this, small amounts of carbon dioxide find uses in such consumer applications as:

– Theater effects
– Night club, theater, and haunted houses smoke and fog machines.
– De-gassing of tanks when such are used for volatile, flammable chemical storage. Other inert gases, such as argon can be used in this application as well.

Carbon dioxide is also used in the production of urea, most of which is used as fertilizer. The Bosch–Meiser process, the reaction chemistry for which is shown in simplified form in Figure 6.14, was developed nearly a century ago, in 1922. It is actually a two-step process in which a carbamate is produced, but never isolated.

Figure 6.14: Urea production.

Nevertheless, current uses of carbon dioxide such as this are not nearly large enough to offset the amounts of carbon dioxide emitted from coal-fired power plants, or from metals refining.

6.4 Inorganic molecules and compounds

As with organic molecules, inorganic materials can exist in all three physical states of matter. Within this large array of matter, several elements can exist as discrete molecules. Again, here we will concentrate mostly on those which are solids, with some exceptions.

6.4.1 Sulfur and sulfur crowns

Sulfur can be found in nature as an element, often in deep deposits. In such cases, it is extracted via the Frasch process, in which a series of pipes is inserted into the deposit, and superheated water is blown in to solvate it and allow pressure to bring the sulfur to the surface. Hence crystalline, elemental sulfur forms eight-atom sulfur crowns, as shown in Figure 6.15.

But sulfur is now often claimed as a by-product in the recovery and purification of crude oils, since government standards have gotten stricter about allowing any sulfur to remain in the refining process, and since corporations have found it more profitable to remove, recover, and use sulfur. In such cases, the sulfur is usually captured as H_2S, and not elemental sulfur.

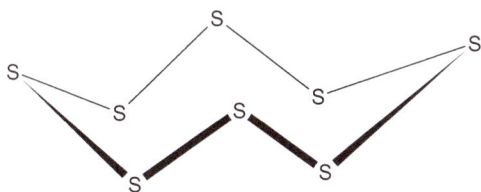

Figure 6.15: Elemental S_8 crowns.

Additionally, sulfur can be obtained through the refining of metal elements from sulfide ores. In the past, roasting of ores had been accompanied by the release of large amounts of sulfur oxides. Increasing and continued environmental awareness, however, has made obvious that such gases are major pollutants; they have been captured at the point of their emission, for later reuse.

Almost all sulfur is used on an industrial scale for the production of sulfuric acid. The production of sulfuric acid is the largest single chemical commodity produced in the world.

The United States Geological Survey Mineral Commodity Summaries 2021 tracks the production of sulfur, and recently reported: "In 2023, recovered elemental sulfur and byproduct sulfuric acid were produced at 86 operations in 26 States." Further, it states: "Elemental sulfur production was estimated to be 8.0 million tons" [3]. The production of sulfur and of sulfuric acid as well is obviously a large, widespread operation. Statements like this also indicate that sulfuric acid production continues to be a major use of elemental sulfur [4].

6.4.2 Sulfuric acid production

Using elemental sulfur as a source, the reaction chemistry illustrating the production of sulfuric acid is illustrated in five steps, as shown in Figure 6.16. Termed the contact process, this is the means by which much of the world's sulfuric acid is produced.

$S + O_{2(g)} \rightarrow SO_{2(g)}$

$SO_{2(g)} + 1/2\, O_{2(g)} \rightarrow SO_{3(g)}$

$SO_{3(g)} + H_2O \rightarrow H_2SO_4$

$H_2SO_4 + SO_{3(g)} \rightarrow H_2S_2O_7$

$H_2S_2O_7 + H_2O \rightarrow 2H_2SO_4$

Figure 6.16: Large-scale sulfuric acid production.

The reason the reaction is not halted at the initial production of H_2SO_4, the third step, is because direct addition of water to sulfur trioxide creates a highly corrosive mist.

Instead, sulfur trioxide ends up being absorbed into the aqueous concentrated acid. This forms what is often called oleum ($H_2S_2O_7$). The fifth step of Figure 6.16 is the final addition of water to oleum, forming concentrated sulfuric acid.

6.4.3 Iodine, I_2

Iodine is the only readily available halogen that exists naturally as a solid, with bromine being a liquid, and chlorine and fluorine both being gases. It is produced mainly from caliche, a sedimentary rock found in different locations worldwide.

6.4.4 Chlorine

Elemental, molecular chlorine is a material that finds extensive use in numerous applications. Its history is decidedly checkered, as it was the first of the poisonous gases deployed during the First World War, near Ypres, in 1915. However, the use of chlorine in drinking water has become one of the most effective and least costly ways to purify and decontaminate water, making it drinkable, and reducing water-borne disease throughout the world.

The Chlor-Alkali process is used to produce this material today. Figure 6.17 shows an idealized schematic of what is called the membrane method of its production. This is not the only means by which chlorine can be produced via electrolysis, but it is the most common of three processes. Note that elemental chlorine is one of three products obtained from this process. Traditionally, sodium hydroxide has been the economic driver for the process.

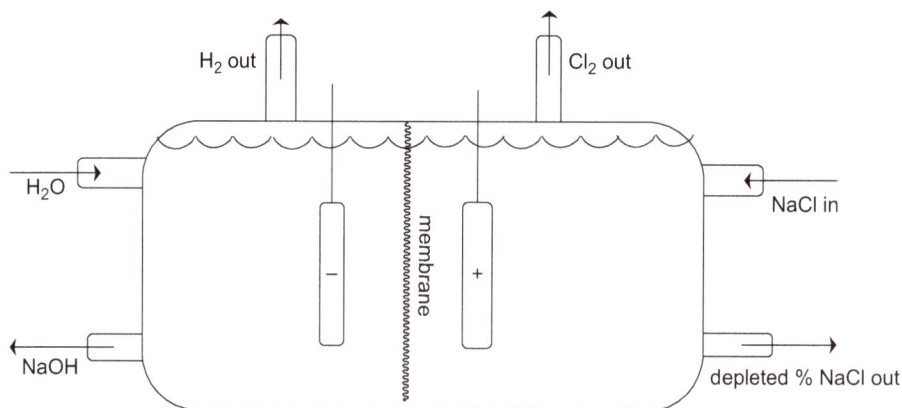

Figure 6.17: Chlor-Alkali process.

6.4.5 Interhalogen compounds

A wide variety of molecules have been produced that are termed "interhalogen" compounds, because they are composed of two different elements from the halogen column of the periodic table. Table 6.2 shows several, although more than 20 are now known.

Table 6.2: Interhalogen compounds.

Name	Formula	b.p. (°C)	Uses	Comments
Chlorine monofluoride	ClF	−100 °C		Lightest interhalogen
Bromine monochloride	BrCl	5 °C		
Iodine monochloride	ICl	97 °C	Halogenating agent	Solid at room temperature
Chlorine trifluoride	ClF_3	12 °C	Reactant in making UF_6 from used uranium	
Bromine trifluoride	BrF_3	126 °C	A fluorinating agent	
Iodine pentafluoride	IF_5	98 °C	Fluorinating agent	Will react with glass
Iodine heptafluoride	IF_7	5 °C	Strong fluorinating agent	

Interhalogen compounds are routinely prepared by the direct combination of the elements. This means that such work must be handled with great care, since elemental fluorine is extremely corrosive, even etching and dissolving glass.

6.4.6 Allotropes of phosphorus

Elemental phosphorus does not occur in nature. When it is isolated, routinely from phosphate ores, it exists in several different allotropes, as discrete molecules, with P_4 and P_{10} being the most common. Figure 6.18 shows the structures of both allotropes.

The bond angle stress in P_4, often known as white phosphorus, is such that when it is exposed to air it combusts violently, with the product being phosphorus pentoxide (P_4O_{10}). The heat created, as well as the smoke created by such a combustion – the latter being mostly aerosolized P_4O_{10} – has made white phosphorus a munition used by many armies in the twentieth century.

The form of phosphorus that looks at first glance like graphite is often called black phosphorus, but does not have a flat structure. Rather, it is referred to as a trun-

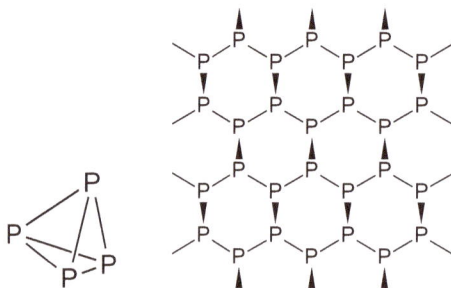

Figure 6.18: Phosphorus allotropes.

cated planar arrangement of atoms. Highly pure black phosphorus has been used to produce thin films through chemical vapor deposition, or CVD.

6.4.7 UF_6

An inorganic molecule that has found a specific use since it was first made in the 1940s is uranium hexfluoride (sometimes simply called "hex"). It has a low boiling material – with a boiling point of 56.5 °C – which has an octahedral structure, as shown in Figure 6.19.

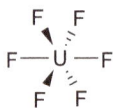

Figure 6.19: Lewis structure of UF_6.

Uranium hexafluoride has found extensive use in the nuclear industry, where it is used in the production of isotopically enriched uranium. In a gaseous state, U-235-F_6 will move faster than U-238-F_6, and thus can be separated via gas centrifuge or gaseous diffusion. The reason this uranium-based gas is used instead of any other is that fluorine is mono-isotopic. This means that the only difference in molecular mass of any sample of UF_6 is the result of differences in the isotopes of uranium.

6.4.8 SF_6

Another inorganic molecule that finds use in the gas state is sulfur hexafluoride. Shown as its Lewis structure in Figure 6.20, sulfur hexafluoride can be generated through the direct combination of the elements.

Sulfur hexafluoride has been found to be a very good dielectric medium or insulating medium, and thousands of tons have been produced annually. Unfortunately, it

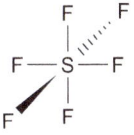

Figure 6.20: Lewis structure of SF_6.

is one of the most powerful of the greenhouse gases, and there are efforts directed at finding substitutes for it in these applications.

6.5 Hybrids, metal-organic frameworks

Metal-organic frameworks (MOFs) represent a form of hybrid between distinct molecules and molecular materials that incorporate some metal-containing portions that are not strictly covalently bonded [5]. Often they are crystalline solids. A large number of MOFs have been constructed since the 1990s, often with the aim of creating a large, yet stable, central space in a cubic structure in which some chemical species can be stored, or specific chemical reactions can take place [5–7]. Figure 6.21 shows the schematic for a simple MOF, noting that corners are often composed of metal cations, and what are sometimes called linkers are organic ligands.

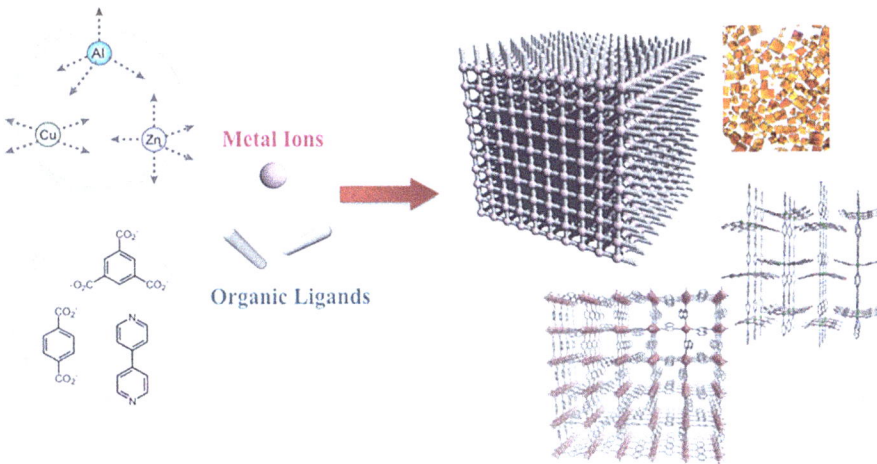

Figure 6.21: Example of MOF. Originally published in [8].

As mentioned, and seen in Figure 6.21, the general idea underlying the structure of any MOF is a cubic arrangement of eight corner groups that are inorganic in their composition. The corners are then connected by twelve spacer groups which are organic and rigid [8]. For this, para-substituted phenyl groups have proven to work well. When connected, the eight corners and twelve linkers form a cubic structure that has a large internal cavity.

The cavity in a MOF is often large enough to store small molecules such as hydrogen or methane, making these materials potentially useful should such molecules be used as a truck or automobile fuel, or used in other energy storage devices [7].

6.6 Recycling possibilities

As with many materials, there are significant possibilities for recycling, depending on the item or material. We have already noted that plastics recycling is a mature industry, one that continues to grow. We have also mentioned that it is far easier to recycle solids than it is liquids, and that it is very difficult to recycle gases.

Some of the molecular materials we have mentioned in this chapter, however, have no real potential for recycling. For example, one of the lighter hydrocarbons, propane, is useful in many applications. One such application is seen in Figure 6.22, propane as an automotive transportation fuel. This means, however, that it is combusted to CO_2 as an end product, and no recycling is possible.

Figure 6.22: A propane-fueled bus.

References

[1]   A.R. Paris, A.B. Bocarsly. High-efficiency conversion of CO_2 to oxalate in water is possible using a Cr-Ga oxide electrocatalyst, *ACS Catalysis*, 2019, 9(3): 2324–2333.

[2]   L.A.J. Garvie. Decay of cacti and carbon cycling, *Naturwissenschaften*, 2006, 93: 114–118.

[3]   United States Geological Survey, Mineral Commodity Summaries, 2024. Downloadable.

[4]   Sulfuric Acid Today. Website. (Accessed 16 June 2025, as: https://h2so4today.com/).

[5]   V. Blay, L.F. Bobadilla, A.Z. Cabrera. Metal-Organic Frameworks, 2018, Walter DeGruyter GmbH. ISBN: 978-904853671-9.

[6]   H.-C. Zhou, J.R. Long, O.M. Yaghi. Introduction to metal-organic frameworks, *Chemical Reviews*, 2012, 112: 673–674.

[7]   L.E. Mphuthi, E. Erasmus, E.H.G. Langner. Metal exchange of ZIF-8 and ZIF-67 nanoparticles with Fe (II) for enhanced photocatalytic performance, *ACS Omega*, 2021, 6(47): 31632–31645.

Chapter 7
Polymers

7.1 Introduction and history

Prior to the twentieth century, the number of plastics or polymers used by human-kind was much more limited than today, simply because of the source materials and refining capabilities that are available today. Table 7.1 provides a list of those which were used by people long before molecular structure or bonding was understood. This list may not be exhaustive, but does cover the major natural polymers and examples of their uses.

Table 7.1: Naturally occurring polymers.

Name	Common use	Source(s)	Comments
Cellulose	Wood and paper	Plants	Cotton is a cellulose form traditionally used in making sails
Proteins	Leathers, clothing	Animals	Animal skins require chemical, preservative treatment to become leathers
Rubber	Tires	*Hevea brasiliensis*	Isoprene units, routinely vulcanized after 1844
Silk	Clothing	*Bombyx mori* – silkworm	Considered a luxury in many parts of the world
Wool	Fibers for clothing, tents, other	Sheep, alpacas, goats, musk oxen, rabbits	Made into cloth worldwide

The common uses for the naturally occurring polymers in Table 7.1 are not the sole uses that these materials have seen throughout history, since some have grown with time, while others have faded. For example, untreated rubber did not have a particularly large market, although after vulcanization in 1844 the use of rubber increased enormously. As a counterexample, leather armor and leather firewood carriers were used extensively in the past when woven cloth was expensive, time consuming to produce, and valuable enough that snags or pulls were to be avoided at all costs, but are seldom seen today. Wool was used extensively for clothing in the past, because it keeps a person warm even when wet. While its market share has decreased because of the use of synthetic fibers for clothing, wool items can still be purchased today.

It can be argued that the Second World War produced advances in plastics and polymers that are still with us today, or that the war became the change agent for industries that did not exist prior to that conflict. Some examples include:

https://doi.org/10.1515/9783112205822-007

7.1.1 Nylon – as a replacement for silk

As the war commenced, it is true that DuPont already produced nylon largely for stockings, and thus for consumer use. In 1942 that production had changed radically, with nylon now being used to make parachutes as well as material cording used in tires. The ability to produce nylon on such a large scale meant that when the war was over, either production would have to be decreased – along with a loss of revenue – or other uses besides military applications would have to be found. Today, nylon is used for much more than articles of clothing, and finds use as a component in tires, as well as in various injection molded parts.

Additionally, it is fair to say that hundreds of derivatives of the first nylon have been produced since the Second World War, with the results being a wide variety of different materials, often with properties aimed at a specific application or use. The original nylon is nylon-6,6, and the Lewis structure for which is shown in Figure 7.1.

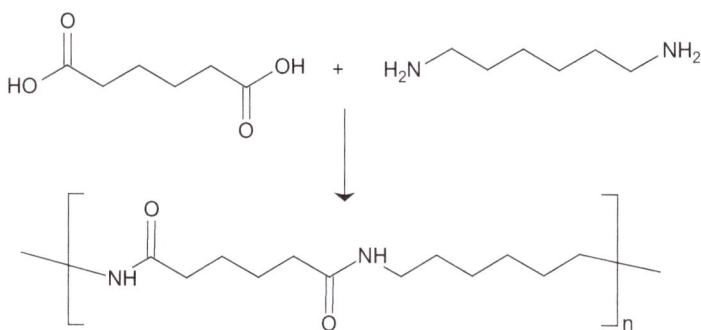

Figure 7.1: Repeat unit, Lewis structure of nylon-6,6.

7.1.2 Synthetic rubber – starting with 1,3-butadiene

The *Hevea brasiliensis* tree grows naturally only in the Amazon basin, and because of the value of the latex extracted from it, export of the tree or its seeds were jealously guarded by the Brazilian government well into the 1800s. Eventually, seeds were secretly smuggled out of Brazil by British adventurer and entrepreneur, Henry Wickham, decades prior to the Second World War. Because Wickham was able to accomplish this, rubber trees were later cultivated in several other tropical lands, many of them colonies of Great Britain when the seeds and trees were first introduced.

Importantly, by the outbreak of the Second World War, much of the rubber produced worldwide had been produced from trees grown from these smuggled seeds in areas that were either directly controlled by the Japanese Empire, or in close proximity to it. Thus, some form of substitute for the latex produced by the trees was needed. The 1,3-butadiene molecule is one that could be isolated and produced from crude oil, and that could be polymerized to produce a polymer that was structurally close to natural rubber. Figure 7.2 shows isoprene and the polymerization of it to natural rubber. The molecule 1,3-butadiene simply lacks the methyl group (shown at the top of the structures in Figure 7.2).

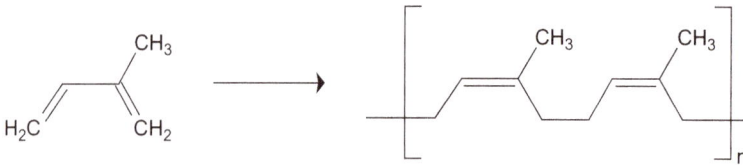

Figure 7.2: Lewis structure of polymerized 1,3-butadiene.

7.1.3 Naugahyde – PVC-coated knit fiber, as artificial leather

Trademarked in 1936, the initial production of Naugahyde was the work of Byron Hunter, at United States Rubber Company. It is a woven fabric as a backing, and polyvinyl chloride (often PVC) as a coating. The material made from this became an inexpensive alternate to leather, and within a short time, a few years, became a well-known material to the general public. Its advantages over natural leather are that: (1) it is inexpensive, (2) it is easy to clean, and (3) it can be produced in long rolls, negating any seams in an end object, which are often necessary when leather from animal hides is utilized. It continues to be used in applications as widely differing as shoes and furniture.

7.1.4 Polytetrafluoroethylene (PTFE) – generally called Teflon

PTFE was discovered by Roy J. Plunkett at DuPont, the general reaction chemistry of which is seen in Figure 7.3. Once again, the discovery was prior to the Second World War, but it was considered expensive, and not marketed. The production was increased vastly when it was found that this new material was unreactive to UF_6 gas, which needed to be handled as part of the isotopic enrichment efforts of uranium for the production of the first atomic weapons, in what is called the Manhattan Project. Yes, this ultimately led to Teflon® pans and nonreactive coatings for other surfaces, and in 1969 also to Gortex®, a water-repellant fabric composed of PTFE that had been stretched and spun into fibers. This material is now used extensively in outdoor wear and raincoats. In addition, numerous variations on this polymer have been created, routinely in the search for materials with specific properties. The ability to repel water or oils is important in many applications, and is a feature of most fluorinated polymers.

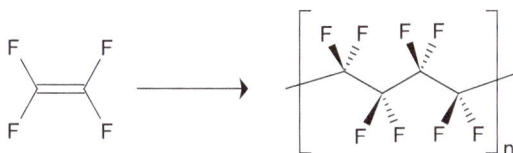

Figure 7.3: Schematic production of and Lewis structure of Teflon®.

The influence that modern, synthetic plastics have had on the world's markets, and on the production of consumer products since the Second World War, on the quality of life that people live, is absolutely enormous. This is an entire class of materials that hardly existed at all before the 1940s, but that has become absolutely essential to a wide number of fields since then. Choosing just one example, the idea of running a hospital without plastics and polymers today is absolutely unthinkable. Their presence helps raise the quality of health care that people receive, and enhance our standard of living.

Yet with the expansion of plastics into every area comes the problem of their long-term disposal. Many of these materials were made and marketed in the early 1950s as materials that would last "forever." This becomes a problem when a consumer decides to discard some plastic item, for whatever reason. The degradation and decomposition of items that have been designed to last "forever" – and that are often designed not to react with moisture or soils – is a challenge that had not previously been thought of. The sheer number of plastic items, or items with plastic components in them, that ended up in landfills became a cause for alarm starting in the 1960s, and continuing to this day.

7.2 Petroleum-based polymers

The vast majority of plastics in use today are petroleum-based. The production and isolation of vast amounts of ethylene and propylene during the distillation of crude oil means that much of the chemistry of polymer production is the chemistry of the double bond. Additionally, the production of styrene and vinyl chloride has become large industries, and finds use almost exclusively in the production of polystyrene and polyvinyl chloride respectively.

There are many polymers that have been produced and marketed for a specific purpose, but six have been so widely applied that what are called Resin Identification Codes are now designated for each of them. Importantly, these six materials have been the focus of numerous efforts to find some means of recycling that is economically feasible, simply because these plastics are so unreactive and long-lasting. There are numerous national and regional organizations exclusively concerned with plastics uses and recycling. These organizations are also concerned with the general public learning something of the lifecycle of polymeric materials, because raising awareness among the general public is believed to be directly linked to greater efforts at recycling [1–16].

The RIC of the six major polymers are shown in Table 7.2.

Table 7.2: The six major polymers, names, RICs, and examples.

RIC	Name	Common abbreviation	Uses, Examples
1	Polyethylene terephthalate	PETE, PET	Soft drink and water bottles
2	High-density polyethylene	HDPE	Bottles, wear- and corrosion-resistant piping, plastic lumber
3	Polyvinylchloride	PVC	"Plastic" piping, cabling insulation
4	Low-density polyethylene	LDPE	Plastic grocery and garbage bags
5	Polypropylene	PP	Food containers, living (single piece) hinges
6	Polystyrene	PS	Packaging material, Styrofoam® cups

The monomer for plastics with RIC codes 1–6 are shown in Figure 7.4. Note that only PETE is produced from two monomers. HDPE, PVC, LDPE, PP, and PS are all made from a single monomer, and all depend upon the ability of the double bond to open and bond to other, like molecules. Also note that RIC 2 and 4 have the same starting monomer. Their reaction conditions as well as additives determine whether they will be high or low density as finished plastics.

There is an RIC of 7, which in the United States is used for plastic materials that are not those made of the first six, or made of blended plastics with two or more polymers in them, or plastic materials that have been recycled at least one time [17].

The existing system of RIC numbers was developed by the Society of the Plastics Industry in 1988. The goal of this was and remains ease of recycling plastic materials used in consumer packaging through some form of pre-recycling separation. The system now seems to be followed by many nations [4, 5].

Beyond just these codes, there are what can be called subsets of various polymers. For example, polypropylene can be produced with all the side-chain methyl groups on the same side of the main chain, or with those methyl groups alternating from one side to the other, or with the methyl groups having no pattern to them at all. This type of three-dimensional arrangement is called tacticity. Such arrangements as these three make significant differences for the finished polymer. For example, isotactic polypropylene, the repeat unit of which is shown in Figure 7.5, is the type made and marketed on a large scale. It has a rigid enough structure that it is useful in many applications. Syndiotactic polypropylene also has some rigidity and exists as a solid,

terephthalic acid and ethylene glycol	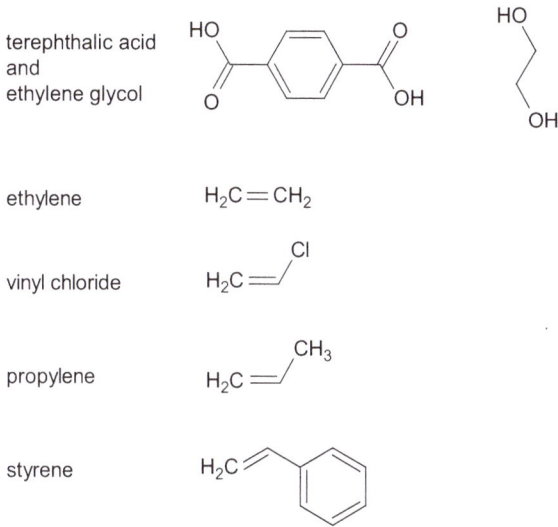

ethylene	$H_2C = CH_2$
vinyl chloride	$H_2C = \overset{Cl}{\diagup}$
propylene	$H_2C = \overset{CH_3}{\diagup}$
styrene	$H_2C \diagup \bigcirc$

Figure 7.4: Lewis structures of the starting monomers for plastics with RIC 1–6.

with the repeat unit that is shown in Figure 7.6. Atactic polypropylene is essentially a viscous liquid or semi-solid with only limited uses, and its lack of order is difficult to represent with a simple Lewis structure.

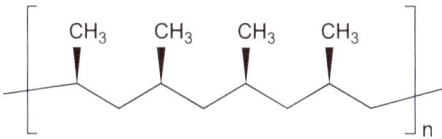

Figure 7.5: Isotactic polypropylene.

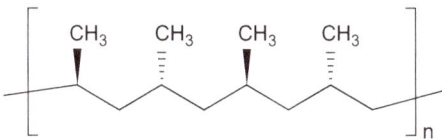

Figure 7.6: Syndiotactic polypropylene.

As might be expected, the production of the different types of polypropylene is dependent upon reaction conditions, and especially on the choice of catalysts used to promote the polymerizations. What are called Ziegler-Natta catalysts have been extensively studied and tried, in order to find those that produce each type of polypropylene exclusively, and to find which do so with the greatest efficiency – what is often called the highest turnover.

The production of the various types of polyethylene (HDPE or LDPE), polyvinyl chloride (PVC), and polystyrene (PS) is also affected by the choice of catalyst and reaction conditions. PVC and PS can also be produced in the three tacticities just mentioned. Interestingly, it is atactic polystyrene that is the most useful, and that is made into many consumer end products. Polystyrenes of more ordered tacticities tend to be too brittle and difficult to work [17].

7.2.1 Other petroleum-based polymers

The six polymers for which RICs have been assigned are certainly not the only plastics that have been used on a large scale. A nonexhaustive list of others is found in Table 7.3, along with their raw materials and example uses.

Table 7.3: Petroleum-based polymers without RICs.

Name	Raw material	Example uses
Acrylic (polymethyl methacrylate, PMMA)	Methylacrylic acid	Optical lenses
Acrylonitrile-butadiene-styrene	Mixture of three monomers: styrene, acrylonitrile, polybutadiene	Automotive parts
Bakelite (polyoxybenzylmethylen glycolanhydride)	Phenol, formaldehyde	Kitchen items, insulators
Nylon (PA, polyamide)	Adipic acid and 1,6-diaminohexane	Clothing, rope, molded parts
Polycarbonate (PC)	Bisphenol-A, phosgene	Compact discs, DVDs, building materials
Polyetherimide (PEI)	Bisphenol-A dianhydride, m-phenylene diamine	Medical equipment
Polyimide	Pyromellitic anhydride, 4,4'-oxydianiline	Heat-resistant applications
Polylactic acid (PLA)	Lactic acid	Filaments for 3-D printing (now a bio-based plastic as well)
Polyoxymethylene (POM, aka acetal)	Anhydrous formaldehyde	Gears, because of low coefficient of friction, and high strength
Polysulfone (PSU)	4-chlorophenylsulfone, diphenoxide	Any application where toughness is needed
Polyurethane (PU)	Diisocyanate, diol	Wide range of applications, since numerous diisocyanates/diols can be used

These plastics and many others have found uses either in industry, or in some consumer product area. In cases where the size of the user end item is small, or their incorporation into a larger device or piece of equipment is required, some are seldom recycled. In many cases, the RIC of "7" is applied to such plastics, which simply indicates that they cannot be recycled as one of the six plastics for which individual RICs have been assigned [17]. Figure 7.7 shows one example of this on a consumer end use item, a food container.

Figure 7.7: RIC code of 7 on a bottle.

7.2.2 Plasticizers

Although the six plastics with RIC codes are made on an industrial scale throughout the world, and although several other polymers, such as those shown in Table 7.3, are made in large quantities, their properties can be tailored to vary widely through the addition of various plasticizers [18]. Such materials are added to the base plastic in what is essentially a trial-and-error method, until a final material is attained with the desired properties for some application – usually involving enhanced flexibility without any loss of durability. Indeed, it is fair to say that there is an entire plasticizer industry within the plastics industry, because the plasticizers must themselves be made in enormous quantities. Figure 7.8 shows the Lewis structure of one of the most common plasticizers.

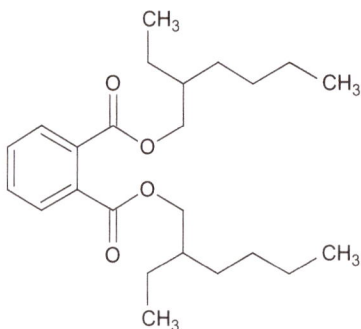

Figure 7.8: Lewis structure of bis(2-ethylhexyl)phthalate.

7.2.3 Recycling of petroleum-based polymers

The recycling of polymers with RIC codes 1–6 has become an established industry in most countries. By far, the easiest and most profitable of the six major plastics to recycle is RIC 1, PETE (polyethylene terephthalate). This is because of its density, its ability to be cleaned after an initial use (think of cleansing beverage bottles after they have been emptied), and its ability to retake a form after melting. Plastics such as RIC 6, PS, and polystyrene are recycled much less. This is because its density is so low – certainly when blown through air during formation of an object – that the costs for shipping and re-forming it are prohibitively high.

Figure 7.9 shows a typical recycling station in a grocery store. These are convenient places for them, although such bins and machines are now also located in busi-

Figure 7.9: Plastics recycling point.

nesses and corporations, and are sometimes simply waste bins which are emptied by some employee, then taken to more established spots.

7.3 Renewable and nonpetroleum-based polymers

The durability of the plastics produced in the largest amounts, as well as the use of fossil fuels to produce their monomers, means that all of the RIC1-6 plastics have been examined to see if attempts can be made to produce any of them from bio-based, renewable materials, or to produce materials that have the same basic properties of these six plastics. Often, the problem evolves to one of making a plastic that feels and performs the same as one of the established six, but that has a renewable source, usually plant-based, even if the chemical composition of the plastic produced is not the same chemical as one of the six.

The challenge of producing plastics in what is called a circular economy – one that does not use fossil fuel starting materials and that produces no long-term waste – is threefold, as follows:

1. Find a source of monomer that is renewable, either from an animal source or from a plant source.
2. Produce plastics that can be recycled without downcycling, or that decompose into some mixture of usable biomass, water, and carbon dioxide.
3. Meet the first two challenges in a manner that is economically competitive with the existing production of plastics with RIC 1–6.

Reflecting on these challenges makes one realize the scale of the overall undertaking, and the work that still needs to be undertaken to meet all three goals [19–23].

7.3.1 Polylactic acid (PLA)

Arguably the most successful bioplastic that has been marketed is polylactic acid, or PLA. Biologically, the production of the lactic acid from which PLA is produced is through a fermentation process. Most carbohydrates – C5 and C6 sugars – can be used as the carbon source, and are usually obtained from milk. Several types of Lactobacillus bacteria will produce the lactic acid. From this point the lactic acid is polymerized. As shown in Figure 7.10, PLA is a polyester and also forms as a condensation polymer, releasing a mole of water for each polymer link formed.

PLA certainly meets the challenge of a bio-based source for its starting material. It does degrade, but does not do so in traditional landfills. It requires specialized, monitored composting for it to degrade to any significant extent.

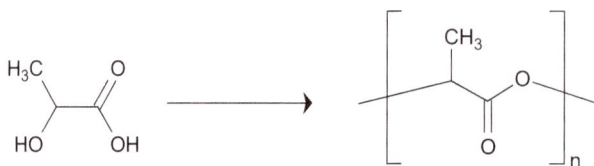

Figure 7.10: Production of polylactic acid (PLA).

7.3.2 Polyhydroxyalkanoates (PHAs)

Another bio-based plastic that has found use are the polyhydroxyalkanoates, or PHAs. Various microorganisms such as *Bacillus subtilis* or *Cupriavidus necator*, or *Alcaligenes latus* are able to ferment sugars or some lipids resulting in an entire class of these polyesters. Figure 7.11 shows the repeat unit of one. We will point out that differences in chain length and in the makeup of the side chain (shown here as a methyl group) result in changes in the physical properties of the polymer produced.

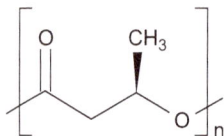

Figure 7.11: Repeat unit of a polyhydroxyalkanoate.

Certain polyhydroxyalkanoates have found uses in medical applications because of their biodegradability. Such uses include items and materials that are used in the human body, such as sutures and surgical meshes. They have not yet been made on a scale to compete with those of RIC 1–6.

In general, the idea of using a variety of plants to produce plastics is laudable in that large-scale use will uncouple any resulting plastics from the finite, nonrenewable resource that is crude oil. The following questions, however, are always part of the larger equation of how to transition to a plant-sourced monomer, or plant-based plastic:

1. What plant species provide the most raw material or monomer per plant?
2. How much water is required for a cycle of plant growth, from seed to harvest?
3. What amount of land will be required for large scale production of a plastic?
4. Will this plant growth and the land required compete with the growth of others plants which are established sources of food?
5. What energy input is involved in harvesting, gathering, concentrating, and processing such plant matter?
6. What costs are involved in transport of plant matter or refined monomer material to the locale or factory of use, where plastics will be manufactured?

These questions must be adequately addressed as such production is scaled up, since all plastics must compete with each other in the world's commodity markets.

7.3.3 Silicones

Carbon appears to be the only element that bonds to itself in long enough linear chains of covalent bonds of only one element, for which the end result is called plastic. But a combination of silicon and oxygen covalent bonding in a repeat unit, and a linear fashion, results in another broad class of plastics, called silicones, or sometimes polysiloxanes [24].

Much like the carbon-based polymers made on a large scale, silicone has a history that does not stretch back too far before the Second World War.

The basic repeat unit of a silicone is shown in Figure 7.12. Note that the main chain of the polymer is a repeating series of silicon atoms, each bonded to two oxygen atoms. Also note that the two remaining bonds to silicon are shown with an "R" group. This is because many different silicones with a wide variety of properties can be manufactured by varying these pendant groups. They can be alkyl organic moieties, can be aryl moieties such as phenyl groups, and even can be substances that cross-link one silicone chain to another.

Because of this wide variety of side chains which can be a component of a silicone, these materials exist as everything from viscous oils to rubbery plastic solids.

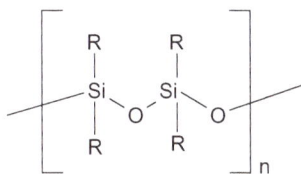

Figure 7.12: Repeat unit of a silicone.

There are other factors that determine the physical properties of the silicones as well. Although the backbone of such polymers is consistently a repeat unit of silicon-to-oxygen atom linkages, and although the R groups, the side chains, strongly influence what the bulk physical properties of a resulting silicone can be, variations in the components of the chain, as well as the length of the chains, and the amount of cross-linking between chains will also play some role in determining the material characteristics of the silicone. For example, polydimethylsiloxane (oftentimes abbreviated PDMS, or dimethicone), shown in Figure 7.13, is one of the most commonly produced silicones. PDMS exists as a clear oil at ambient temperature. Another example is diphenyl-dimethyl-silicone, the Lewis structure of which is seen in Figure 7.14, which exists as an oil that is both odorless and colorless, and that is thermally stable at temperatures as high as 250 °C.

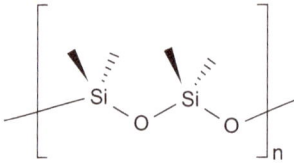

Figure 7.13: Polydimethylsiloxane.

Because silicones have become such a large field, there are now recognized divisions among them. Table 7.4 shows this, but is not an exhaustive list of examples and uses.

Table 7.4: Types of silicones.

Abbreviation	Name	Characteristics	Uses
RTV-1	Room-temperature vulcanization – one-component	Polydimethylsiloxane and cross-linkers	Able to wet most surfaces
RTV-2	Room-temperature vulcanization – two-component	Linear silicone, cross-linkers, and catalysts	Preferred for mold making, and for sheathing electric parts
LR	Liquid rubber	Often cured by peroxide	Usually two-component
HTV	High-temperature vulcanizing	Can be cured at 200–250 °C	Flexible solids, medical applications

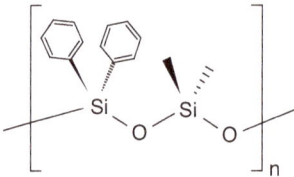

Figure 7.14: Diphenyl-dimethyl-silicone.

7.3.4 Recycling, silicon-based polymers

Much like carbon-based polymers, silicones can be recycled, and in theory can be recycled an infinite number of times. As with most materials, solid silicones are easier to recycle than liquids. In practice, most communities do not tend to have facilities for this type of recycling, and used silicones are often discarded with other household wastes. Some communities do, however, have specific drop-off points for a wide variety of hazardous materials, or materials that are not normally recycled, and silicones can be taken to such locations.

7.4 Polymer composites

As the name implies, polymer composites are plastics produced by the mixing of two or more materials, often with some material – possibly a plastic, possibly some other material – embedded in a polymer. The reason for this is that the material placed into the polymer in some way reinforces it, making the resulting material stronger or in some other way better suited to some particular purpose. What can be called the additive or reinforcing material can be virtually any other substance, such as metals or other polymers, a ceramic, glass fibers, or carbon fibers.

The automobile industry has made extensive use of polymer composites because the composites are routinely stronger than a polymer alone, but are of lighter weight than a metal might be. Lighter weight in an automobile translates into better gas mileage, and thus is of constant interest to this industry.

As one might imagine, the use of polymer composites is also widespread in the aerospace industry. Planes and rockets require strong, yet lightweight, materials for the same reason as the automobile industry. In addition, any spacecraft must have surfaces strong enough to withstand the stresses and pressures of the launch from ground level through the entire atmosphere, ultimately into space. Table 7.5 shows a very brief set of examples of polymer composites. Their total number and their specific uses are very broad.

Table 7.5: Examples of polymer composites.

Name	Composite component	Example use
Polyamides	Carbon fibers	High strength, lightweight needs
Polyesters	Glass or carbon fibers	Airplane or automotive parts
Polypropylene	Inorganic fillers	Automotive parts
Polypropylene	Wood fibers	Weathering-resistant decking

7.5 Thermoset polymers

We have been classifying polymers by RIC, and when possible by specific chemical formulas. But all polymer materials can be divided into what are called thermoset polymers and thermoplastic polymers.

When a polymer is considered a thermoset it simply means whatever material is made from it irreversibly forms the final product. Often pressure or heat is applied to produce the thermoset, and in many cases both. A wide number of thermosets have found a specific use in one or more industries. Common and well-known examples are included in Table 7.6. Once again, this table represents only a small sampling of all the possible uses of these materials.

Table 7.6: Examples of thermosets and their uses.

Thermoset	Possible uses
Bakelite	non-conducting items
Duroplast	Items used in clean or sterile environments
Melamine resin	Consumer end use objects, e.g., cookware
Epoxy resin	Adhesives
Polyisocyanurate	Building insulation
Polyester fiberglass	Insulators
Polyimide	Electronics
Polyurethane	Adhesives and coatings, insulating foam
Urea-formaldehyde resin	Particle board composites and wood adhesive
Vulcanized rubber	Tires, durable fiber

Thermosets are molded and formed through several different processes. They include:

1. Compression molding. As the name implies, this requires the use of increased pressure. It has been used extensively in making a wide variety of items, by forcing the material into a desired shape.
2. Extrusion molding. Forcing the thermoset plastic through an extruder while the plastic is warm has proven to be an excellent means by which piping and other items can be formed. A variety of shapes can be produced using different extruder heads.
3. Injection molding. Much like extrusion molding, this process involves pushing elastic material that has been heated into some mold of a desired shape.
4. Spin casting. Once again, heated material is used, in this technique being pushed into a desired shape via centrifugal force. After cooling, the shape is retained.

The major difference between thermosets and thermoplastics (to be discussed next) is that thermosets are formed irreversibly. Because of this it is difficult to recycle objects that are made of thermosets into some other form or other object. Shredding or mechanical degradation of the thermoset material has been tried as a means to find some second use for thermosets.

We should mention, since we have treated natural rubber to some extent in this chapter, it is worth noting that rubber is one of the toughest thermosets known. This becomes a problem when people try to find a use for old tires, which often accumulate otherwise. But two niche uses for shredded rubber are as matting or ground cover in children's playgrounds, and as molded bricks used as walkways for some of the elite horse farms of the world.

7.6 Thermoplastic polymers

Polymer materials that soften above some temperature, and harden or become glass-like below that temperature, and that are recyclable, are referred to as thermoplastic polymers. This temperature at which the material changes is termed the glass transition temperature, and is routinely given the symbol T_g, It is specific to each different polymer, and polymer blends and composites sometimes have very broad ranges for their T_g. While numerous plastics have some specific T_g, some simply melt without undergoing this softening (such a temperature is generally designated T_m, the melting temperature). Table 7.7 lists the very common plastics we have discussed. Numerous others exist, often for a specific niche in terms of an item manufactured for some specific end application or consumer use.

Table 7.7: T_g or T_m of the common polymers and blends.

RIC	Name	Abbrev.	T_g (°C)	T_m (°C)
1	Polyethylene terephthalate	PETE	75	255
3	Polyvinyl chloride	PVC	80	
5	Polypropylene	PP		160
6	Polystyrene	PS		240
	Acrylonitrile butadiene styrene	ABS	105	
	Styrene–acrylonitrile	SAN	115	

7.7 Recycling

The recycling of plastics is a mature industry that has become extremely large over the past four decades. For example, roughly 1.5 billion pounds of PETE alone are recycled in the United States in just one year. This is based on both the materials properties of the plastics produced on a large scale and the economic ease and profitability of their second and subsequent uses through recovery, remelting, and re-forming.

In theory, plastics can be recycled an infinite number of times. In reality, each cycle of recycling presents the possibility of some contaminant entering the polymer batch, making the next material one that is slightly worse in some characteristic or property. Color is actually a prime example of this. Clear beverage bottles, when recycled, must have any colored labels, or colored cap rings, removed if the recycled plastic is to remain clear, and thus usable as another clear beverage bottle. If this does not happen, the recycled material will have some color to it. This does not make it useless. But it may end up being unacceptable as a beverage bottle a second time. It may end up being used in some application like a rug, or in some other colored material made from spun polymer threads.

Figure 7.15 shows bottles that are ready for recycling. Labels are still on because many stores only take the bottles they sell, and determine which these are via the bar code on the bottle. The bar code must be present when the consumer puts the bottle into the recycling machine. Still, it should be noted that soft drink beverage bottles are routinely recycled in many nations, and are routinely made of PETE, with the RIC of 1. Figure 7.9 showed an example of a recycling station that accepts both plastic and metal. It can accept this wide an input of material because it functions by reading the bar code on each plastic bottle or metal can. This increases the number of items that are recycled, but also connects the recycling to the colored plastic on which bar codes are printed.

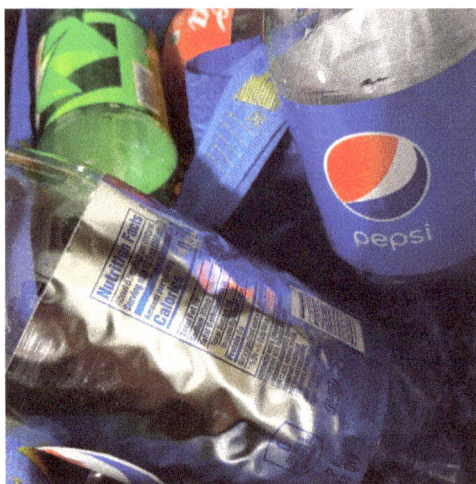

Figure 7.15: Bottles to be recycled.

References

[1]   American Chemistry Council. Website. (Accessed 9 June 2025, as: https://www.americanchemis try.com).
[2]   Plastics Industry Association. Website. (Accessed 9 June 2025, as: https://www.plasticsindustry.org).
[3]   Alliance to End Plastic Waste. Website. (Accessed 9 June 2025, as: https://www.endplasticwaste.org).
[4]   Society of Plastics Engineers. Website. (Accessed 9 June 2025, as: https://www.4spe.org).
[5]   Flexible Packaging Association. Website. (Accessed 9 June 2025, as: https://www.flexpack.org).
[6]   VinylPlus. Website. (Accessed 9 June 2025, as: https://www.vinylplus.eu).
[7]   Chemistry Industry Association of Canada. Website. (Accessed 9 June 2025, as: http://www.plas tics.ca).
[8]   Plastics Europe. Website. (Accessed 9 June 2025, as: https://plasticseurope.org).
[9]   British Plastics Federation. Website. (Accessed 9 June 2025, as: https://www.bpf.co.uk).
[10] IK Industrievereinigung Kunststoffverpackungen. Website. (Accessed 9 June 2025, as: https://kunst stoffverpackungen.de).

[11]   International Council of Chemical Associations (ICCA). Website. (Accessed 9 June 2025, as: https:// icca-chem.org).

[12]   Plastics Industry Manufacturers of Australia. Website. (Accessed 9 June 2025, as: https://www.pima. asn.au).

[13]   The Japan Plastics Industry Federation. Website. (Accessed 9 June 2025, as: https://www.jpif.gr.jp).

[14]   Japan BioPlastics Association. Website. (Accessed 9 June 2025, as: http://www.jbpaweb.net).

[15]   South Korean Manufacturers of Plastic. Website. (Accessed 9 June 2025, as: https://panjiva.com).

[16]   Plastics SA. Website. (Accessed 9 June 2025, as: https://www.plasticsinfo.co.za).

[17]   M. Elzageheid. Macromolecular Chemistry: Natural and Synthetic Polymers, 2022, Walter DeGruyter GmbH. ISBN: 978-311076276-1.

[18]   Plasticisers.org. Website. (Accessed 9 June 2025, as: https://www.plasticisers.org).

[19]   A. Ammala. An overview of degradable and biodegradable polyolefins, *Progress in Polymer Science*, 2011, 36(8): 1015–1043.

[20]   T. Narancic, F. Cerrone, N. Beagan, K.E. O'Connor. Recent advances in bioplastics: Application and biodegradation, *Polymers (Basel)*, 2020, 15 Apr (online). doi: 10.3390/polym12040920

[21]   R. Luque, C.-P. Xu. Biomaterials, 2016, Walter DeGruyter GmbH. ISBN: 978-311034230-7.

[22]   E.S. Stevens. Green Plastics: An Introduction to the New Science of Biodegradable Plastics, 2002, Walter DeGruyter GmbH. ISBN: 978-069121417-7.

[23]   K. Ghosh, B.H. Jones. Roadmap to biodegradable plastics – current state and research needs, *ACS Sustainable Chemistry Engineering*, 2021, 9(18): 6170–6187.

[24]   J.J. Chrusciel. Silicon-based Polymers and Materials, 2022, Walter DeGruyter GmbH. ISBN: 978-311064367-1.

Chapter 8
Metals and metallurgy

8.1 Introduction

The use of metals profoundly shapes our lives, and has done so for almost all societies for millennia. Ages of several civilizations are often named after bronze or iron, with bronze routinely coming before iron (or steel), since this copper–tin alloy has the lower melting point and working temperature. Both metals though are important in the development of several areas of a society, such as the ability to grow food with better tools, to work other materials, and to produce weapons. As global exploration expanded, and people who lived in remote areas for millennia were brought into contact with other peoples, there are known cases in which entire peoples moved so that they could be in closer proximity to a people who used and traded metals tools and objects.

Several metals have been known since ancient times, but the discovery of many more in the eighteenth and nineteenth centuries has continued to change and shape aspects of the global society today, broadly expanding the possible number of usable alloys. Indeed, there are now so many alloys that standards have been created to help define and categorize them [1].

8.2 Elemental metals

The periodic table of the elements is roughly three quarters metals, although some of them occur extremely rarely in nature – such as iridium – others are only produced in nuclear reactions – such as americium – and some never occur in nature by themselves – such as gallium or potassium. While seven metals were known from ancient times, a much larger number were discovered in the eighteenth and nineteenth centuries.

8.2.1 Metals from antiquity

Any discovery of the following metals has been lost to time, but the metals listed here in Table 8.1 have been known since ancient times, and have often been associated with a body in our solar system. They include the following:

The alchemists used specific symbols for these metals – and planets – as a means of identifying them, but also as a means of keeping this knowledge in closely guarded circles of people they deemed worthy of such.

https://doi.org/10.1515/9783112205822-008

Table 8.1: Metals known in ancient times.

Element	Symbol	Traditional name	Astronomical association
Mercury	Hg	Hydrargentum	Mercury
Copper	Cu	Cuprum	Venus
Silver	Ag	Argentum	The Moon
Gold	Au	Aureum	The Sun
Iron	Fe	Ferrum	Mars
Tin	Sn	Stannum	Jupiter
Lead	Pb	Plumbum	Saturn

Gold, silver, and copper, arranged in column IX of the transition metals of the periodic table of the elements – and known for centuries as the coinage metals – all have ancient histories. All are mentioned very early in the Bible and in other major religious texts. Copper even takes its name from, or gives its name to, the island of Cyprus, where the metal was heavily mined in ancient days. Copper was used extensively to make tools and weapons, and modern archaeologists continue to excavate ancient copper mines at places like Timna, in southern Israel [2]. Silver and gold were more often used for valuable and decorative objects, or as adornment both for persons and for objects of worship. Indeed, of all the metals on the periodic table, only gold and copper are elemental metals that are not "silver" in color.

Other metals were also known in antiquity. Lead was found to be easy to work, to have a lower metaling point than metals such as iron and copper, and was used extensively in the Roman Empire. It found use in building projects, and in lining aqueduct channels in some places. The words "plumber" and "plumbing" have their origins in the Latin word for lead – plumbum.

Iron was also known in ancient times. Some of the earliest iron objects found as grave goods in Egyptian tombs were iron beads and knives. Since Egypt has no iron deposits, these objects are believed to have been forged from meteoric iron. The rather colorful name for the knives found in these tombs is: daggers from heaven.

Tin was likewise known in ancient times, although large deposits of it are widely scattered. However, the idea of mixing it with copper to make a metal stronger than either alone appears to have arisen in more than one place, and to have spread widely, as tin was traded between long distances, such as the Balkans and Italy, at or before the time of the Roman Empire.

Mercury was also known in ancient times, and acquired a name that described its behavior: quicksilver. Ancient scholars had some debate as to what this material was, since it looked like and had the density of a metal, yet flowed like water. The atomic symbol Hg comes from the Latin: hydragyrum or hydrargentum, meaning "water" and "silver." But the element was also known in ancient China, and appears in the tomb of the famous emperor Qin Shi Huang, who commissioned the army of terra cotta warriors to be stationed in his tomb [3].

8.2.2 Iron

A significant number of large, multinational companies produce both iron and steel – a class and series of alloys based on iron, usually including up to 4% carbon in the iron. The top manufacturers of iron and steel are shown in Table 8.2, but there are many more, and the positions can change from year to year, depending on markets and outputs.

Table 8.2: Largest iron and steel corporations [4–8].

No.	Name	Location	Mass (tons)	Other metal products
1	Vale	Brazil	300 M	Nonferrous metals
2	Rio Tinto	UK-Australia	286 M	
3	BHP	Australia	248 M	Copper
4	Fortescue Metals Group	Australia	204 M	
5	Anglo American	UK headquarters	61 M	

The United States Geologic Survey (USGS) tracks the annual production of iron nationally in the following different forms: iron, steel, iron ore, plus iron oxide pigments [9].

Tracking iron production does not necessarily correlate with the production of steel, since ore and iron can be shipped to furnaces and metal refineries across national borders. Besides the USGS Mineral Commodity Summaries, iron and steel production is tracked by other organizations as well, including: the World Steel Association, Eurofer, the Japan Iron and Steel Association as well as other professional trade organizations [10–15]. Additionally, corporate web sites advertise their production capabilities, usually emphasizing the quality and value of their products [16–38].

8.2.2.1 Ore sources

Globally, there are a variety of iron ores which can be mined. Table 8.3 is a nonexhaustive list of the major ones. It is noteworthy that the majority of iron is refined from three ores: hematite, magnetite, and taconite. These three generally possess high iron percentages.

It is worth mentioning that prior to the Second World War, taconite ores were not mined extensively. This is because hematite and magnetite ores were readily available. When these sources were exhausted and played out, taconite became a profitable iron source.

Iron production from virtually all different ores is always a series of chemical reductions of some starting iron compound, often an oxide such as taconite. The process ends in the reduced iron metal. This occurs along with the removal of oxygen via a reducing agent, usually carbon monoxide. This means a source of carbon, such as

Table 8.3: Iron ores.

Name	Formula	Iron %	Geographic Location	Possible Other Metals
Ankerite	$Ca(Mg,Mn,Fe)(CO_3)_2$	Varies	Peru	Magnesium, manganese
Goethite	$FeO(OH)$	62.8		
Greenalite	$Fe_4Si_2O_5(OH)_4$	52.3	USA, Minnesota	Mixed oxidation state iron
Grunerite	$Fe_7Si_8O_{22}(OH)_2$	39.1	South Africa	
Hematite	Fe_2O_3	69.9		
Laterite	Mixed $Fe_xAl_yO_z$	Varies	India, Australia,	Aluminum, nickel
Limonite	$FeO(OH){\cdot}nH_2O$	52.3		
Magnetite	Fe_3O_4	72.4		
Minnesotaite	$(Fe,Mg)_3Si_4O_{10}(OH)_2$	30.7	USA, Minnesota	
Siderite	$FeCO_3$	48.2		
Taconite	Fe_3O_4 mixed with quartz	*Usually >15	USA, Minnesota, Michigan	Silica

*Within taconite, iron is often present as magnetite in some dispersed form.

coke, is essential to the refining operation. The simplified reaction chemistry for this is shown in Figure 8.1. It begins with iron in a high oxidation state:

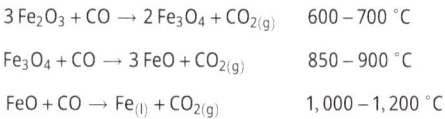

$$3\,Fe_2O_3 + CO \rightarrow 2\,Fe_3O_4 + CO_{2(g)} \qquad 600 - 700\ ^{\circ}C$$

$$Fe_3O_4 + CO \rightarrow 3\,FeO + CO_{2(g)} \qquad 850 - 900\ ^{\circ}C$$

$$FeO + CO \rightarrow Fe_{(l)} + CO_{2(g)} \qquad 1{,}000 - 1{,}200\ ^{\circ}C$$

Figure 8.1: Iron production reaction chemistry.

The reducing agent in each step is carbon monoxide, which ultimately is sourced from coke, and which is oxidized to CO_2. The coke must be combusted with air blasts. For this, the basic chemical reaction is

$$2C + O_{2(g)} \rightarrow 2\,CO_{(g)}\ 200 - 700\ ^{\circ}C$$

Limestone is also used when reducing iron ores, because all such raw ores do contain some silicate impurities. In all cases, limestone is calcined (highly heated) to produce calcium oxide. This is then combined with the silicates in the ore to produce a coproduct referred to as slag.

A curious indicator of the size of the iron and steel industry is that a National Slag Association exists, a trade organization that attempts to find productive uses for this coproduct material [39]. The production from limestone of calcium silicate, slag's primary component, is as follows:

$$CaCO_{3(s)} \rightarrow CaO + CO_{2(g)}$$

$$CaO + SiO_2 \rightarrow CaSiO_3$$

Enormous blast furnaces are required for the large scale production of iron metal. Currently, almost fifty companies produce steel. One furnace is capable of producing over 80,000 short tons of iron every week. Figure 8.2 shows a basic blast furnace design.

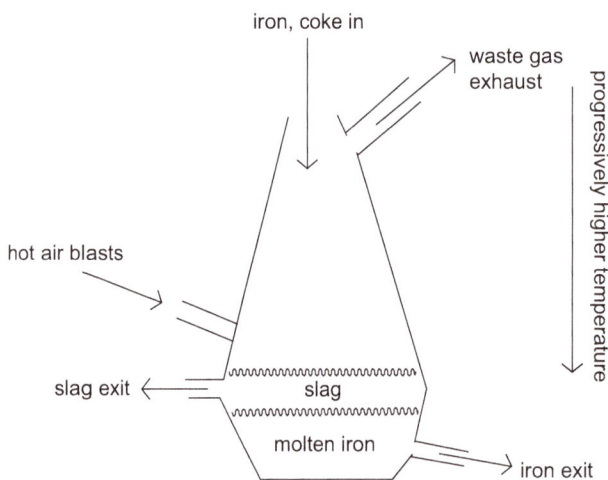

Figure 8.2: Blast furnace.

Any blast furnace has to be lined with a refractory material – brick, basically – and have some entrance port or ports for the iron ore, limestone, and coke near or at the top. The bottom is where heated air is forced up through the furnace; and it may require hours for any solid material that has been added to the top to descend to the bottom and through the increasingly hotter zones. Through this time, the starting iron ore is chemically reduced to molten iron, with liquid slag being produced also. At regular intervals, both must be drained from the furnace.

Blast furnace sizes are not easy to grasp from photographs, and certainly not from diagrams. Routinely, however, such an operation is large enough that any decision to commission or decommission the furnace is a corporate one. Such are made after intense study of the economics of the operation at a company's highest levels. Any blast furnace functions for years or decades when it becomes operational, and only maintenance, or some type of an accident stops its functioning.

8.2.2.2 Steel production

Close to 99% of all iron that is refined is further worked into steel. The term actually includes a large number of alloys, not all of which are solely iron and carbon. Steel is so common that organizations exist to track its uses, and to advocate for it, such as the World Steel Organization. This organization claims that more than 3,500 steel formulations are in use, although some are for admittedly small, very specific uses. Table 8.4 is a very abbreviated list, focusing on common steels used in large volumes.

Table 8.4: Common steel types.

Steel type	Alloyed with	Uses or physical properties
Carbon steel	C	High hardness
High-speed steel	W	Enhanced hardness
High-strength low-alloy steel	1.5% Mn	Enhanced strength
Low alloy	Mn, Cr, <10%	Enhanced hardness
Manganese steel	Mn, 12%	Wear resistance/durability
Stainless steel	Cr 11%, Ni	Corrosion resistance
Tool steel	W, Co	Enhanced hardness, drills and cutting tools

Of note, virtually all alloying elements are used to make the resulting material harder than the starting elemental iron. This seems to be a rare phenomenon, one associated with iron. It is difficult to find other elemental metals that are made harder when they are made impure – when they are alloyed.

Steel is routinely produced in either a basic oxygen furnace, usually abbreviated BOF, or an electric arc furnace – EAF – in large amounts. Figure 8.3 shows a schematic of a basic oxygen furnace.

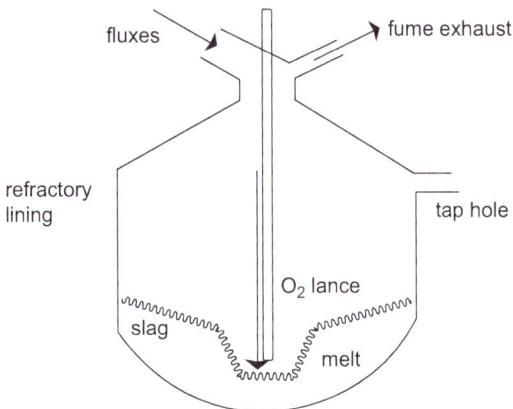

Figure 8.3: Basic oxygen furnace.

The heart of a BOF is the oxygen lance. The introduction of elemental oxygen at high temperature and high flow forces some of the carbon in the metal to oxidize to carbon dioxide. This produces high-quality, low-carbon steel. The reaction is a high-temperature process which requires that the shell of the furnace is made of refractory material, with the container being made of steel.

The electric arc furnace is the newest means by which steel can be produced on a large scale, although the first patents for such a furnace were taken out in the 1880s. At this early time, such furnaces were not necessarily used for iron and steel production. The Second World War marked the time at which electric arc furnaces were most extensively used for the production of steel. Figure 8.4 shows the basic schematic of such a furnace. Note that the electrodes in the center of the apparatus are the major difference between an EAF and a BOC. Similarities include inlet and outlet spouts for flux and for slag, as well as the main container being made of a refractory material. Since EAFs work on a batch-by-batch basis, some series of levers are always incorporated into the base of the furnace, so that molten product can be tipped out.

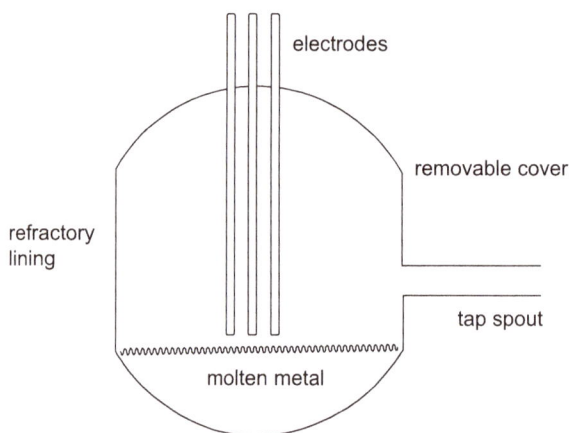

Figure 8.4: Electric arc furnace.

One difference between the two types of furnaces involves the input. Scrap iron can be used when priming an electric arc furnace. Traditionally, iron ore, as well as coke and limestone has been used for the BOF.

8.2.2.3 Historical applications of steels

Historically, several civilizations have gone through what is called an "Iron Age," based on this metal. Less well known, but still commanding a certain amount of interest, are a few types of steel that have been produced on a small scale, at least when compared to production today. The following three are famous examples of steel and its uses.

1. Damascus steel – or Wootz Steel

The attractive, whirled patterns seen in some medieval swords and knives are often called Damascus steel because this is the city in which they were made. The name also has become associated with high quality, and indeed, many of these weapons remain uncorroded after centuries (although, admittedly, they have been cared for). Modern analysis has found that these weapons have approximately 0.083% vanadium in them. Curiously, the iron ingots or pigs from which they were forged appear to have come from India, where it was called Wootz steel. The "secret" of making such steel was lost for centuries, probably because of a disruption of trade routes between the two locales. The chemical composition has been rediscovered recently [40], and thus this type of steel can again be forged.

2. Toledo steel

Toledo, Spain is another city that has a long history associated in part with its fame in making steel. It was noted as far back as the wars of Rome that the swordsmiths of this city produced blades of unusual strength. The strength was due to the construction of such swords using two different steels, with different carbon contents. One was harder than the other, and when hot welded together, the resulting weapon was both strong, and yet had some flexibility – meaning it was not brittle. Toledo steel was apparently used widely among the Roman legions when it could be obtained.

3. The Japanese katana

The long Japanese sword called a katana was the larger of two swords carried by the traditional samurai warriors. Like the previous two, this type of steel has over time acquired a reputation that is almost mythical. The production of such swords was a time-consuming, intense process, often requiring months for a single blade. During forging, clay was applied to one edge of the blade, making it heat and cool at a different rate that the opposing edge, the end result being the gentle curve associated with such swords. The end result was a cutting edge that was extremely sharp and hard, and that held that edge very well. Over 100 of these swords from past periods in Japanese history have been declared national treasures of Japan.

Ceremonial swords are still part of many military dress uniforms today, often for commissioned or noncommissioned officers. This custom hearkens back to times when swords were a primary weapon for soldiers, even though the use and improvement of firearms over the past few centuries has made swords an essentially obsolete weapon. The three examples just given still have a place in history; but as far as materials chemistry is concerned, they have never been made on a large scale.

8.2.2.4 Uses of iron and steel

The USGS's annual Mineral Commodity Summaries tracks iron and steel production because it is vital to the U.S. economy, and because the variety of uses for it is enormous. It is fair to say that the extensive use we make of iron and steel make much of modern life possible. Recently, the Mineral Commodity Summaries commented on uses for the metal: "Construction accounted for an estimated 30% of net shipments by market classification, followed by service centers, 24%; automotive, 14%; converting and uses, 8%; non-classified shipments, 4%; machinery and equipment, 3%; appliances, 3%; and other applications, 12%" [9]. Some of these categories are obvious to the general public, although other, often niche uses are not. Beyond the USGS there are several organizations, both regional and national, which act as advocates for the uses of steel and iron. These organizations often make and update lists of uses at their websites [10–15, 25–37].

Concerning iron, the USGS Mineral Commodity Summaries makes a comparison between it and competing materials when discussing possible substitute materials for iron. It states: "Iron is the least expensive and most widely used metal. In most applications, iron and steel compete either with less expensive nonmetallic materials or with more expensive materials that have a performance advantage. Iron and steel compete with lighter materials, such as aluminum and plastics, in the motor vehicle industry; aluminum, concrete, and wood in construction; and aluminum, glass, paper, and plastics in containers" [9]. We discuss both aluminum and plastics in other parts of this book, but will note here that decisions on whether or not to use iron versus one of these other two materials, both of which are less dense, often becomes a decision between the overall mass of some finished product – such as an automobile – and the strength of that object or end user item.

8.2.2.5 By-product production

The refining of iron from ores always has two major by-products associated with it: gaseous carbon oxides and some amount of solid slag. Slag is relatively easy to recover, and thus a trade organization has been formed which advocates for productive uses for slag. Such uses are not normally seen by the general public – such as the base material in road beds. As well, slag can be mixed into cement formulations, resulting in different types of concrete [39].

While slag can be considered some type of coproduct of iron production, since it can be used, the carbon oxides – carbon monoxide and dioxide – cannot, since they are gases and difficult to recover on a large scale. Some economically feasible form of carbon capture or sequestration remains a goal of the iron and steel industry.

The following is an examination of the production of iron that one might see in a freshmen-level college chemistry course. It shows the production of one ton of iron, then the amount of carbon oxides that are coproduced. We do this here to present some scale as to how much of this gaseous by-product is formed when iron is refined. Figure 8.5 illustrates the reaction.

$$Fe_2O_{3(s)} + 3\,CO_{(g)} \longrightarrow 2\,Fe_{(l)} + 3\,CO_{2(g)}$$

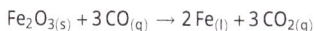

Figure 8.5: Iron reduction from ore to metal.

Beginning with one ton of iron metal we can compute he mass of carbon dioxide that is coproduced as a by-product:

$$1\text{ metric ton }Fe_2O_3 = 2,000\text{ pounds} \times 453.59\text{ g/L lb} \qquad = 907,180\text{ g }Fe_2O_3$$
$$907,180\text{ g }Fe_2O_3 \times 1\text{ mol }Fe_2O_3/159.687\text{ g} \qquad = 5,681.0\text{ mol }Fe_2O_3$$
$$5,681.0\text{ mol }Fe_2O_3 \ \times 3\text{ mol }CO_2/1\text{ mol }Fe_2O_3 \qquad = 17,043\text{ mol }CO_2$$
$$17,043\text{ mol }CO_2 \times 44.010\text{ g }CO_2/1\text{ mol }CO_2 \qquad = 750,061\text{ g }CO_2$$

Reconverting to tons:

$$750,061\text{ g }CO_2 \times 1\text{ lb}/453.59\text{ g} \qquad = 1,653.6\text{ lb }CO_2$$
$$1,653.6\text{ lb} \times 1\text{ ton}/2,000\text{ lb} \qquad = 0.827\text{ ton }CO_2$$

We see here then that for each ton of iron that is refined and reduced from ore, over 80% of a ton of CO_2 is also produced. Based on the enormous tonnage of iron that is manufactured annually, it seems obvious that one or more practical means to capture and use the CO_2 which is coproduced is extremely important.

8.2.2.6 Iron recycling

Scrapyards in many areas have been devoted to iron and steel recycling because the economic advantages of the process are so great. In some operations, like the scrapping of an ocean-going ship, the used metal to be recycled never arrives at a scrapyard. In such cases, the metal goes directly to a furnace for remelting and re-forming. An army of workers cut such ships apart, and the metal that made them up is entirely remelted and reused.

Scrap yards exist in all nations with some large-scale manufacturing capability, in which iron and steel is recycled. The stimulus is routinely economic, since it is much cheaper to recycle the metal than to produce new metal from its ores. Recycling of iron and steel is a large enough and mature enough industry that trade associations exist devoted to it [37, 38]. This is the same situation for several other metals, and thus steel and iron recycling is often part of some larger scrapyard facility.

8.2.3 Silver

8.2.3.1 Isolation and refining

Silver has been a valuable metal since antiquity, and thus people have sought it out many times while exploring parts of the world. The largest transfer of silver the world has probably ever seen is that of the Spanish moving it from their colonies in the New World – lands that include modern-day Mexico and Bolivia – back to Spain, over the course of centuries.

Although silver has at times been discovered in concentrated, localized deposits, many times it exists in low concentrations in some larger geologic formation. In most cases, the silver exists as some sulfide mineral. This means it must be concentrated prior to refinement, and other elements must be separated. What is called the Washoe process for silver concentration and purification was first used while the Comstock Lode of present-day Nevada was initially being worked. This is actually an adaptation of a significantly older, considerably slower process called either the Patio process or Cazo process. This process was used at Cerro Rico in Bolivia, for centuries, and was used elsewhere, when silver had to be concentrated.

The Washoe process does not work in precise, stoichiometric ratios, but the general steps can be summarized as follows:
1. Sizing. Ore samples are crushed to a grain size roughly that of sand or possibly smaller.
2. 1,200 pound batches are loaded into copper pans, then made into a slurry.
3. An amalgam is created by the addition of mercury.
4. The liquid at this point is predominantly liquid. Both copper sulfate and sodium chloride are then added.
5. The batch is agitated and stirred using iron paddles or plates.
6. Steam is utilized to heat this slurry.
7. Impurities in the batch are separated, with only a silver–mercury amalgam remaining.

At this point in the process the mercury must be driven off. Perhaps obviously, this is the environmental concern, and this is the problem with this process that makes it unfavorable today. With the advent of inexpensive electricity, the Washoe process is now considered obsolete. Silver is currently refined by a combination of smelting and leaching processes, after which electrolytic refining takes place. Silver-containing ore batches, especially copper ores which have some silver as a coproduct, the following steps, in some combination, are required to extract and purify the silver:
1. Mechanical and physical crushing and milling. This maximizes the surface area of the material and creates uniform sized particles.
2. Separation via froth flotation. Various surfactants are mixed with the powdered material to aid in separation of the silver minerals from any others based on different hydrophobicities. The separation permits silver-bearing ore to be enhanced in concentration.
3. Smelting. Especially with copper ore batches, first smelting the overall sulfide, then electrolytic refining, further concentrates the silver from any batch as an anode mud, sometimes also called anode slime.
4. Anode mud smelting. This is done to oxidize any other metals present except the silver, plus gold and what are called platinum group metals (often abbreviated

PGM). At this point, the metal is concentrated sufficiently that relatively pure ingots can be produced.

5. Electrolysis. Silver ingots undergo electrolysis at this point, in a solution bath of silver nitrate as well as copper nitrate. The final product is high-purity silver.

8.2.3.2 The Parkes process

First patented in 1850, the Parkes process, named after inventor Alexander Parkes, is a means of separating silver from lead metal. This is done by mixing molten zinc with molten lead. Although the two metals are essentially immiscible in each other, silver is several thousand times more miscible in zinc than in lead. The result is that any silver in the lead, and any gold that might be present as well, migrates to the molten zinc. The molten zinc–silver alloy is then heated until the zinc is vaporized. The remaining silver is then captured. The zinc can then be recaptured and reused. Figure 8.6 illustrates the basic reaction chemistry of the process.

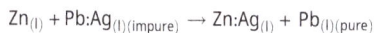

$$Zn_{(l)} + Pb{:}Ag_{(l)(impure)} \rightarrow Zn{:}Ag_{(l)} + Pb_{(l)(pure)}$$

Figure 8.6: Parkes process.

8.2.4 Gold

The history of gold might be considered a large part of the history of the human desire for wealth. There is no other metal that attracts the attention of people as much as gold. It has an ancient history, and almost all cultures and civilizations that work metal have found gold to be beautiful as a form of ornamentation. Elemental gold can be found in quartz veins, as well as in river beds, the latter as what are called placer nuggets. These placer nuggets and other easily gathered gold has been that which was used by most ancient civilizations.

The origin of coinage is connected with an early use of gold. Nuggets of gold, or of a naturally occurring alloy of gold and silver that has been called electrum, were found in the rivers of what is now Turkey. To ensure some standard to their value, they were punched with some image, which eventually became a standard image for an area, or a city-state. The nuggets stamped with a lion or bull were associated with the city-state of Lydia, and became the world's first coins.

As a material, gold was found to be extremely useful by many ancient civilizations, because it is highly malleable and ductile. Jewelry and other forms of ornamentation could be made from gold simply because it is very easy to deform, either in cold working, or when heated.

Gold and its value has also been one of the reasons that many people have moved from their homes and explored and settled in some other region. The myth of the fabled city of El Dorado brought waves of Spanish to the New World. Gold rushes in the

United States and Canada prompted the movement of people first to the southern Appalachians, then to the North American west coast, then to the Yukon Territory. In the twentieth century, the desire for gold on the part of a single government – that of the USSR – was the reason tens of thousands of people were sentenced to years (and often to their deaths) in distant Siberia. None of this was driven by the material properties of gold, such as the just mentioned ductility or malleability, or even its excellent electrical conductivity. Rather, these voluntary or forced migrations were driven by the perceived value of this yellow metal.

Curiously, despite several industrial uses for gold as a metallic material, people have again moved to regions of California to search for gold that was overlooked in the previous rush [41]. It is still a matter of the wealth associated with the metal, which trades at approximately $3,200 per ounce in 2025.

8.2.4.1 Refining and isolation

Because of the value of gold – it qualifies as a high-value, low-volume commodity more than many other metals – numerous companies exist that are in some way involved with its isolation, refining, and purification. For example, at the beginning of the year 2022, the value of gold has been at or over $1,900 per troy ounce. Table 8.5 is a nonexhaustive list of companies that are involved with gold. In almost all cases, each of these companies also is invested in other metals, including copper and silver, and including PGM. In fact, several of the companies, as well as others not listed in Table 8.5, have a primary product that is not gold. Rather, gold is a coproduct of some other mining and refining operation. Numerous organizations, either governmental or nongovernmental, track the production of gold, and often advocate for its use or ownership [9, 42–44].

Table 8.5: Companies that produce gold (and other metals).

No.	Company name	HQ location	Other products	
1	Barrick Gold Corp.	Canada	Silver	[6]
2	Newmont Mining Corp.	USA	Copper, silver	[7]
3	AngloGold Ashanti	South Africa	Silver	[8]
4	Gold Fields, Ltd.	South Africa	Copper, molybdenum, platinum, palladium, nickel, silver	[9]
5	Newcrest Mining, Ltd.	Australia	Copper	[10]
6	Kinross	Canada	Silver	[11]

Table 8.5 (continued)

No.	Company name	HQ location	Other products	
7	Goldcorp Inc.	Canada	Silver	[12]
8	Yamana Gold	Canada	Copper, silver	[13]
9	Agnico-Eagle Mines	Canada	Silver, copper, zinc	[14]
10	Polyus Gold	Russia		[15]

Throughout history, gold nuggets have been found either on the Earth surface, or in streams and rivers, or in deposits that are very close to the surface and do not require extensive mining equipment. Large nuggets have a certain allure and charm associated with them, and have often been named. Table 8.6 lists several of them.

Table 8.6: Named, famous gold nuggets.

Name	Weight (oz.t.)	Weight (lb.)	Location of find	Find date – fate
Alaska Centennial	294.1		Ruby, Alaska	1998
Armstrong nugget	80.4		Grant County, Oregon, USA	1913
Beyers and Holtermann	≈5,000	630	Hill End, Bathhurst, Australia	1872
Boot of Cortez	389		Mexico	1989
Butte nugget	70	6.07	Butte County, California, USA	July 2014, sold to private collector
Dogtown nugget		54	California, USA	1859
Fricot nugget	201	6.25	California Gold Rush	At California State Mining & Mineral Museum
Golden Eagle	1,136		Western Australia	1931
Hand of Faith	875	27	Kingower, Victoria, Australia	1980
Heron	1,008	69.08	Mt. Alexander goldfield	1855
Highland Centennial	27.495		Montana	1989
Ironstone's Crown Jewel	528		Jamestown, California	1992

Table 8.6 (continued)

Name	Weight (oz.t.)	Weight (lb.)	Location of find	Find date – fate
Lady Hotham		98.5	Ballarat, Australia	September 8, 1854
Mojave Gold Nugget	156		Randsburg, California	1977
Normandy nugget	177		Ballarat, Australia	
Pepita Canaa	1,951	133.80	Serra Pelada Mine, Para, Brazil	September 13, 1983, now at Banco Central Museum
Welcome Nugget	2,218	152.1	Bakery Hill, Ballarat, Australia	1858, melted in 1859
Welcome Stranger	2,520	173	Moliagul, Victoria, Australia	1869

Even though any individual gold nugget can be considered large, together they are only a very small part of the annual production of gold, and its uses, which are: "Estimated global gold consumption, excluding exchange-traded funds and other similar investments, was in jewelry, 46%; central banksand other institutions, 23%; physical bars, 16%; official coins and medals and imitation coins, 9%; electrical and electronics, 5%; and other, 1%" according to the USGS Mineral Commodity Summaries [9]. A statement like this provides information on how much gold is not really used at all, but rather kept as some form of a store of wealth.

The extraction of gold from ores is never a matter of stoichiometrically balanced chemistry, but a relatively simple reaction is shown in Figure 8.7.

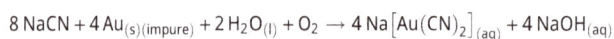

$$8\,NaCN + 4\,Au_{(s)(impure)} + 2\,H_2O_{(l)} + O_2 \rightarrow 4\,Na\left[Au(CN)_2\right]_{(aq)} + 4\,NaOH_{(aq)}$$

Figure 8.7: Isolation of gold as a complex.

This reaction is a technique titled either the cyanide process, or the MacArthur-Forrest process. It isolates gold from other elements as a cyanide complex, which means some further chemical reactions are required to reduce the gold back to a solid metal. What is termed the carbon in pulp process (the CIP process) accomplishes this, again in a nonstoichiometric fashion. The following steps accomplish this:

1. Addition of carbon particles – these and the gold-bearing complex are washed together.
2. Separation and removal – the mixture of carbon particles and gold complex are is removed from the solution and again washed.
3. Elevated temperature and basic pH – this step removes the gold cyanide from the overall solution.

4. Electro-winning – this electrochemical process isolates solid gold metal upon the cathode.
5. Final smelting – gold-bearing cathodes are then smelted. This forms them into ingots that can be further used

When copper or silver is present in a gold-bearing batch, this process works does not work as well. However, the process has become more common in the past three decades. It is effective should an ore require concentration from some percentage that is very low and not economically workable, to one that is.

8.2.4.2 Anode muds

The rather dull term "anode mud" or "anode slime" is used to describe the tailings of an electrochemical process by which one metal is separated from another. Copper refining is often accompanied by the coproduction of anode muds that may contain gold, silver, or the less common PGM. Nearly 10% of domestic gold in the United States is captured in these muds [9].

8.2.4.3 The Miller process

When chlorine gas is directed over a batch of metal that is predominantly gold, metal chlorides are formed with those elements other than gold. The solid salts – metal chlorides – are not soluble in gold, and can thus be separated from the molten gold. This technique can produce gold as high in purity as 99.9%. Figure 8.8 is an attempt to show this in terms of reaction chemistry.

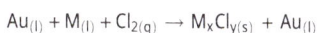

$$Au_{(l)} + M_{(l)} + Cl_{2(g)} \longrightarrow M_xCl_{y(s)} + Au_{(l)}$$

Figure 8.8: Miller process.

8.2.4.4 The Wohlwill process

This process, patented in 1874, can produce gold with a purity of 99.999%. Curiously, it does require a chloroauric acid ($HAuCl_4$) solution which is not extracted or recovered from the solution. What are termed Doré bars – gold anodes – of at least 95% purity are used along with pure gold cathodes as the two electrodes in an electrolysis solution. During electrolysis, gold accumulates at the cathode. When the mass is sufficient, the cathode is removed and a new starter cathode is placed. The cathode which is removed is routinely 99.999% pure.

8.2.4.5 The gold carat system – 18, 14, 12, 10 carat

Because such a large amount of gold is used as a store of wealth and as ornamentation, we will look briefly at a system for defining gold purity that has centuries of history to it, the carat system. Even though percentages may be more convenient, the carat system remains very common.

Carat – also spelled karat – is a means of dividing the purity of gold into 24 parts. Pure gold is 24-carat gold, while 22-carat gold has been used in the past for certain coins, like the British sovereign. This 22 parts is 91.67% gold, the remainder usually being copper. Two other popular carat weights, at least in terms of jewelry, are 18-carat (or 75%) and 14-carat (or 58.33%) gold. Lower carat weights, such as 10 carat, or 41.66%, are also used in some applications.

8.2.4.6 Uses of gold

We have seen, above, the breakdown of the uses of gold, according to the USGS Mineral Commodity Summaries [9], and will note that gold is one of only a few materials that are not actually used as much as they are stored. Gold ingots and what are called gold bullion coins are two ways in which people store wealth. We will though look at the other uses for gold.

8.2.4.7 Gold jewelry

Pure gold is seldom used in jewelry, because the metal is soft enough that it can be scratched with a fingernail. Rather, 18-carat and 14-carat gold are often used, with the remaining parts – to bring it to 24 carats, total – being copper. This makes for a harder material, one that lasts decades or even centuries without showing significant wear.

High school, college, and championship rings are often made of gold. Two companies, Jostens and Balfour, together account for a large percentage of the sales of such rings within the United States. Both companies tend to sell rings that are either 10 or 14 carats in purity. Figure 8.9 shows examples of these.

Figure 8.9: Examples of 10-carat gold college and championship rings.

8.2.4.8 Gold investment coins

The idea of a gold coin used not as circulating money and as a means of exchange, but rather as a way to store wealth, appears to have its origin in modern times in the one-ounce gold coins, called Krugerrands, of South Africa. South Africa had something of a monopoly on such coins until the 1980s, when several other governments began minting their own coins for the purposes of personal investment. Sometimes called "bullion coins" as well as "noncirculating legal tender," the number of nations and territories now issuing such coins has become large. Table 8.7 is a nonexhaustive list of them.

Table 8.7: Gold bullion coins.

Nation	Coin name	Denomination/size
Australia	Nugget	$^1/_{20}$, $^1/_{10}$, ¼, ½, 1 oz, 2 oz, 10 oz, and 1 kg
Austria	Philharmoniker	$^1/_{10}$, ¼, ½, 1 oz
Canada	Maple Leaf	1 g, $^1/_{20}$, $^1/_{10}$, ¼, ½, 1 oz
China	Panda	$^1/_{10}$, ¼, ½, 1 oz
Great Britain	Britannia	$^1/_{20}$, $^1/_{10}$, ¼, ½, 1 oz, 5 oz
Israel	Tower of David	1 oz
Kazakhstan	Golden Irbis	$^1/_{10}$, ¼, ½, 1 oz
Malaysia	Kijang Emas	¼, ½, 1 oz
Mexico	Libertad (Onza)	$^1/_{20}$, $^1/_{10}$, ¼, ½, 1 oz
Poland	Orzel Bielik	$^1/_{10}$, ¼, ½, 1 oz
South Africa	Krugerrand	$^1/_{10}$, ¼, ½, 1 oz
Switzerland	Vreneli	0.1867 oz
USA	Eagle	$^1/_{10}$, ¼, ½, 1 oz

Gold bullion coins represent a relatively new way for small investors to own some of the precious metal, which is why many of these bullion coins are made in small fractions of an ounce (if, for example, gold costs $1,900 per ounce a 1/10th ounce bullion coin will be priced close to $190). Additionally, this form of gold makes it easy to for a person to accumulate it slowly, over the course of years, rather than having to buy a significant amount at one time.

8.2.4.9 Gold in electronics

Gold finds use in electronics largely in the form of contacts, in spots in a circuit where electrical conductivity is crucially important. Because gold is both a very good conductor, and is very resistant to oxidative corrosion, it finds use here. Emphasis is often placed on using minimal amounts of gold for each contact, simply so the price of the device in which it is used does not increase.

8.2.4.10 White gold alloy

The general public knows the term "white gold" as more than one ally. To many, it means gold alloyed with platinum, usually as 18-carat gold. To others it may mean gold alloyed with silver in a similar ration. Metallurgists tend to use the term for 90% gold alloyed with 10% nickel, which gives both the white appearance, and an alloy more resistant to wear or abrasion.

8.2.4.11 Possible substitutes for gold

Because of its cost (approximately $1,900 per troy ounce in early 2022), substitutes for gold are constantly being sought. The USGS Mineral Commodity Summaries [9] has for years made comments such as: "Base metals clad with gold alloys are widely used in electrical and electronic products, and in jewelry to economize on gold; many of these products are continually redesigned to maintain high-utility standards with lower gold content. Generally, palladium, platinum, and silver may substitute for gold" [9].

8.2.4.12 Gold recycling and reuse

Gold recycling has been a highly developed industry for decades. Virtually every grade of gold is recycled. Curiously, the World Gold Council [42] notes a sharp decline in gold recycling in the recent past. This may be in part because of the global COVID pandemic.

8.2.5 Copper

Copper is one of the ancient metals, and its alloys with tin – to make bronze – have been known for millennia. What are called "Bronze Ages" have occurred in different civilizations at different times, and often correspond with an advance of that culture and civilization. Bronze forged in the ancient Middle East, as well as the amazing bronze sculpture and objects made during the Han Dynasty of ancient China, remains examples of how a specific material – in this case the workability and hardness of bronze – can influence an entire society. Examples include the introduction of bronze tools for farming, which allowed a person to farm more land than comparable wooden tools, and the introduction of bronze weapons, which are more effective than weapons made of wood, stone, or bone.

Many companies refine and produce copper today. Table 8.8 is a listing of the top ten, with the understanding that the rankings do shift sometimes from year to year. It should be noted as well that there are several companies which produce copper, but which consider themselves to be producers of some other metal, usually because that other metal is the product of highest value – such as gold or PGM.

The strength, ductility, and malleability of copper mean that it remains economically feasible to mine and refine copper in many places throughout the world, on all

Table 8.8: Top 10 worldwide producers of copper [55–64].

No.	Name	Location	Amt. (thousand metric tons)	Other products
1	Codelco	Chile	1,757	
2	Freeport-McMoRan Copper & Gold Inc.	USA	1,441	Gold
3	BHP Billiton Ltd.	Australia	1,135	Aluminum, iron, manganese, nickel, silver, titanium, uranium
4	Xstrata Plc	Switzerland	907	Coal, ferrochrome, nickel, zinc
5	Rio Tinto Group	UK/ Australia	701	Gold
6	Anglo American Plc	UK	645	Iron, manganese, platinum, diamonds, nickel, phosphates
7	Grupo Mexico	Mexico	598	Gold, silver, molybdenum, zinc, lead
8	Glencore International AG	Switzerland	542	Gold
9	Southern Copper Corp.	USA	487	Silver, zinc
10	KGHM Polska Miedz	Poland	426	Silver

six inhabited continents. In the United States, copper mining continues in Michigan and in several western states.

8.2.5.1 Copper mining, isolation and refining

Although native copper can be found at or near the Earth's surface, much more of it is mined. Currently, the Bingham Canyon Mine in Utah, in the United States is the world's deepest open pit mine, at more than 1 kilometer of depth. Much of what is mined is not reduced metal; rather, it is in the form of an ore. Table 8.9 is a nonexhaustive listing of various copper ores.

It is not difficult to imagine that with a variety of copper ores, there will be a variety of means by which the element is reduced to the metal. That being said, there are several broad steps in copper refining that are basically common to the different ores. They include:

1. Mining and crushing. Ores and mineral deposits are first mined, then physically crushed into smaller easily workable pieces, usually 1–2 cm in size.
2. Grinding. This step reduces overall particle size until the sample is a powder. The purpose is to maximize overall surface area of the batch, and enhances the efficiency of the remaining steps.

Table 8.9: Copper ores.

Name	Formula unit	% Copper	Locations (examples)
Bornite	$2Cu_2S \cdot FeS \cdot CuS$	63.3	Kazakhstan, Congo
Chalcocite	Cu_2S	79.8	Australia
Chalcopyrite	$FeCuS_2$	34.5	USA, Canada, Norway
Covellite	CuS	66.5	USA, Austria, New Zealand
Cuprite	Cu_2O	88.8	Russia, Italy, USA
Malachite	$CuCO_3 \cdot Cu(OH)_2$	57.3	Congo, Russia, USA, Zambia
Tennantite	$Cu_{12}As_4S_{13}$	51.6	Congo, Namibia, USA

3. Concentration and beneficiation. Ore concentrations of copper can be low, thus water is used separate materials that do not contain copper, in the process concentrating the usable ore.
4. Smelting. To drive off the sulfur, sulfide ores must be smelted. Solvent extraction can be employed with oxide ores to concentrate them. This step may need to be performed more than once, but can ultimately produce copper at purity of 99%.
5. Leaching. By using acid solutions to enhance the ore's solubility, copper oxide ores can be solvated into sulfate solutions.
6. Refinement. Refined but impure copper, that which has been smelted can be cast into metal ingots. These are then available to be used as anodes. Using copper solutions the metal can then be electroplated, depositing solid copper.
7. Electrowinning, or electrolytic processing. The just-mentioned copper anodes can be immersed in a process tank, using extremely pure copper rods as cathodes, with an electric current passing through such a bath. The result is the deposition of extremely pure copper onto the cathodes. Along with this is the formation of a by-product that is often called anode mud. Interestingly, this residue may contain gold, silver, PGM, or other rare elements, and because of its value is also recovered. The final product, the highly pure copper cathodes, may be several hundred pounds.

8.2.5.2 Uses of copper

8.2.5.2.1 Piping
Copper piping and tubing is certainly a familiar use of this metal, one that the general public is very much aware of. Indoor residential plumbing is one common, major use, but piping is used in a multitude of other places as well, both indoor and out. For a detailed listing of types of copper tubing, including how it can be divided not only by size but by pressure ratings, the Copper Development Association publishes a downloadable "Copper Tube Handbook" [65].

8.2.5.2.2 Wiring and machinery

Because copper is used in so many applications, copper wire has been made uniform through a series of American Wire Gauges (or AWG). The systems numeric system means that lower numbers are larger wires that have lower electrical resistance. For example, 12AWG is bigger than 14 AWG. To put this in some context, residential buildings in the United States are routinely built with wires spanning from 10AWG to 16AWG.

Beyond electrical wiring in residential and commercial buildings, copper wire finds wide use in making electromagnets. Electromagnets now have a history that is over two centuries old. In that time, the number of uses for them has bloomed. One that is well known to the general public is that of a large, industrial electromagnet used in a scrap yard to lift and separate scrap iron, such as junked automobiles.

8.2.5.2.3 Copper coinage

What might be considered the normal use of copper in coinage is its use in low-denomination pieces, such as U.S. 1-cent coins, or the 1- and 2-euro cent pieces. But copper has been used in coins traditionally, meaning during the times when circulating coins were made of gold, or silver, or copper. For example, from 1838 until 1933, U.S. gold coins were made from an alloy that was 90% gold and 10% copper. As explained above, copper is often alloyed with gold to make the object more durable.

Likewise, U.S. silver coins were for well over a century made of an alloy of 90% silver and 10% copper, again to make the resulting coins more durable. For centuries, two grades of silver have been recognized for higher-value items. Sterling silver is 92.5% silver, and what is sometimes called "coin silver" is 90% silver. This prevented potential counterfeiters from melting silverware to produce counterfeit coins.

A number of different modern, circulating coins continue to be made using copper alloys, even though they retain a white or silver color. For example, the U.S. nickel – the 5-cent piece – is 75% copper and just 25% nickel. In the Euro Zone, the €0.10, €0.20, and €0.50 pieces are "Nordic gold." This is an alloy consisting of 89% copper, 5% zinc, 5% aluminum, plus 1% tin. Figure 8.10 shows several of these modern coins.

8.2.5.2.4 Copper alloys, bronze

Bronze is always some alloy of copper and tin, and today continues to be used in sculptures, and has traditionally found uses in bell making and until the mid-1800s in producing artillery. The production of an artillery pieces was both an art and a science, and was also expensive. It is for this reason that artillery pieces captured by an army were often later used by that same force. As well, naval artillery pieces were salvaged from wrecked ships and used, whenever possible, even if they had to be scavenged from wrecks under water.

Even today, church bells tend to be made of bronze. The reason for this is quite simple: bronze bells produce musical tones that are pleasant to the ear.

Figure 8.10: Modern coins made of copper or copper alloys.

Bronze also has been used extensively in funerary urns and grave "stones" simply because it is resistant to weathering and because with the passing of time such objects take on an attractive toning or patina [66].

Bronze parts are also used selectively for different types of machine parts, very often marine-based, and routinely performance-based. Thus, the choice to use bronze becomes a matter of superior performance balanced against the potential for cost savings.

8.2.5.2.5 Copper alloy: brass

Brass is always an alloy composed of copper and of zinc, but there can be a wide variety of other elements in the mixture. Brass and bronze are terms that have been used somewhat arbitrarily throughout much of recorded history, simple because early metallurgists were not always aware of the "white" metal with which they were working.

Much like the way bronze compositions are determined, brass formulations are almost always based on how well they function in a certain application. For example, what is often called "cartridge brass" takes its name because of its use in making ammunition casings. This brass must eject from the breech of a gun, rifle, or other firearm immediately after discharge, and not expand with the heat produced at that moment.

Table 8.10 displays several common brass formulas, and provides a representative example of their use. The list is not an exhaustive one.

8.2.5.2.6 Further copper alloys

An alloy that contains copper, but that is predominantly tin, is pewter. Through much of history pewter alloys also contained lead. Modern alloys do not use lead because of its toxicity.

A more modern series of copper alloys, one that has found extensive use in aquatic environments, are copper–nickel alloys – predominantly because of their re-

Table 8.10: Brass compositions (empty boxes mean an element is absent from a composition).

Brass name	Cu	Zn	Sn	Fe	Mn	Ni	Al	Pb	Application(s)
Admiralty	69	30	1						Numerous
Aich's alloy	60.66	36.58	1.02	1.74					Seawater environments
Alpha brass	65	<35							Imitation gilding
Beta brass	50–55	45–50							Castable
Cartridge brass	70	30							Ammunition cases
Gilding metal	95	5							Ammunition
High brass	65	35							Rivets, screws
Low brass	80	20							Metal adapters
Manganese brass	70	29			1.3				US golden dollar coins
Muntz metal	60	40		<1					Seawater environments
Nickel-brass	70	24.5				5.5			UK £1 coins
Nordic gold	89	5	1				5		€0.10, €0.20, €0.50 coins
Prince's metal	75	25							Jewelry
Red brass	85	5	5					5	C23000, cast parts
Rivet brass	63	37							
Tombac	85	15							Jewelry
Yellow brass	67	33							Household adornments

sistance to oxidation / corrosion in such environments. The Copper Development Association makes the statement, "Copper–nickel (also known as cupronickel) alloys are widely used for marine applications due to their excellent resistance to seawater corrosion, low macrofouling rates, and good fabricability" [65]. As well, the CDA notes that one of the most widely used of these alloys is 90–10. The numbers mean 90% copper, 10% nickel.

8.2.5.2.7 Copper, bronze, and brass recycling

As with the other metals we have seen, recycling of copper, brass, and bronze, are all well-developed industries. Recycling of copper and alloys based on copper routinely involves some economic incentive. In short, in virtually every case, it is far less expensive to reuse metal that has already been refined than it is to refine and purify new metal from various ores.

8.2.6 Aluminum – Hall–Heroult process

Aluminum is a modern metal, first discovered in 1827. It only became an inexpensive metal in 1887, however, when the Hall–Heroult process was patented. In the years in between, because of the difficulty of reducing aluminum from its parent ore, bauxite, aluminum was valued as a precious metal. In the 1880s the electric dynamo was also patented, which meant that inexpensive, abundant electricity was available – something upon with the Hall–Heroult process depends. After this point in time, aluminum became one of only a few major industrial metals upon which the developed world depends. While it is resistant to oxidation, almost all uses of aluminum are based on its low density. Because of the multitude of uses for it, there are several companies that produce this metal on a huge scale. Table 8.11 lists the top 10 aluminum producers worldwide.

Table 8.11: Top aluminum producers [67–76].

No.	Name	Location	Amt. (M of metric tons)	Other products
1	UC Rusal	Russia	4.173	
2	Alcoa, Inc.	USA	3.742	
3	Aluminum Corp. of China	China	3.502	Copper, gallium, rare earths
4	China Power Investment Corp.	China	2.693	Coal, power generation
5	Rio Tinto Alcan Inc.	Canada	2.174	Power generation
6	Norsk Hydro ASA	Norway	1.985	Alumina
7	China Hongqiao Group Ltd.	China	1.821	
8	Shangdong Weiqiao Aluminum & Power Co.	China	1.715	Power generation
9	Shangdong Xinfa Aluminum & Electricity Group Ltd.	China	1.63	Power generation
10	Dubai Aluminum Co.	UAE	1.42	Water desalination

Even though several of these aluminum producers, as well as several that are not on the list, are international, or do not have their headquarters in the United States, organizations like the USGS Mineral commodity Summaries tracks its production annually [9]. As well, several other national and international trade associations exist that advocate for its many uses [77–86].

8.2.6.1 Aluminum refining and isolation

Bauxite is the ore from which essentially all aluminum is refined, even though other aluminum-bearing ores do exist. In an interesting note, even though aluminum imparts no color to bauxite, it usually occurs as a red stone because of iron impurities within it. Other materials, such as silicates and gallium impurities, also exist in bauxite and are separated during the early steps of the Hall–Heroult process. The resulting concentrated impurities, called "red mud" in the industry, must in some way be used or disposed of. Iron and gallium have been refined from this material in the past.

Trying to write the reaction chemistry of aluminum refining is difficult to do when trying to show stoichiometrically balanced equations. Still, Figure 8.11 attempts to do this.

$$2O^{2-}_{(l)} + C_{(s)} \rightarrow CO_{2(g)} + 4e^-$$

$$Al^{3+}_{(l)} + 3e^- \rightarrow Al_{(l)}$$

The overall oxidation and reduction is:

$$6O^{2-}_{(l)} + 4Al^{3+}_{(l)} 3C_{(s)} \rightarrow 4Al_{(l)} + 3CO_{2(g)}$$

Figure 8.11: Aluminum refining, reaction chemistry.

Refining of aluminum metal is more complex than these reactions indicate, however. A series of steps generally include:
1. Isolation of alumina (Al_2O_3) from raw bauxite by separating the iron ores. This is done with sodium hydroxide – caustic soda – at both elevated pressure and temperature. This step is traditionally called the Bayer Process.
2. Alumina smelting in molten cryolite – the mineral Na_3AlF_6 – which must be performed in cells that are carbon-lined.
3. The carbon electrodes used in the smelting are themselves oxidized and must periodically be replaced.
4. Siphoning off molten aluminum from the cells. The reduced metal is more dense than the cryolite, must be pulled from the lower portion of the cells.

The mineral cryolite was at first mined and purified, but now synthetic cryolite is used in the Hall–Heroult process, as it is economically feasible to do so. While this is not shown in chemical reactions that illustrate aluminum production, there is no other material which works in this process.

Additionally, few processes use as much electricity as the Hall–Heroult process. To give an example, the ArcticEcon website comments: "For the Hall-Hérault process to function, an electric current of low voltage but from 200,000 to 500,000 amperes must pass continuously through each cell. On average it takes about 15.7 kWh of electricity to produce 1 kg of aluminium. This is what makes aluminium smelting such an energy intensive process" [85]. Because of this, aluminum refineries are sited in places such as Iceland, where electricity (from geothermal sources) is inexpensive and readily available.

8.2.6.2 Aluminum uses
Aluminum finds many uses, either as the elemental metal or as a variety of alloys. The common factor or desired characteristic in most items and objects made of aluminum or an aluminum alloy is the combination of corrosion resistance and light weight. With a density of 2.70 g/cc, aluminum is not the least dense of all the metals.

To compare, beryllium has a density of 1.85 g/cc and lithium has a density of only 0.534 g/cc. But neither of these exists in large enough quantities that they could replace aluminum in any serious manner; and lithium is far too reactive as a reduced metal.

Numerous aluminum-magnesium alloys have been made and find some industrial use. Magnesium is another lightweight metal (with a density of 1.74 g/cc), and when alloyed with aluminum provides added strength to the alloy while it still retains the overall ductility of aluminum. So many lightweight alloys have been made that several organizations exist to ensure their standardization, and that a four-digit code is used to represent them [77–83]. In the system, the first digit indicates the major element – aluminum – and the following indicate alloying elements. Therefore, the 1,000 series indicates nearly pure aluminum. As another example, the alloy 6061 is common, includes both silicon and magnesium, and is composed of 95.85% aluminum.

Aluminum beverage cans are undoubtedly the most common aluminum item the general public thinks of when considering into what objects aluminum can be made. As well, aluminum and aluminum alloys, such as aluminum-titanium alloys, are widely known to be used in aircraft and the aerospace industry. But numerous other applications are known. Alcoa – short for the Aluminum Company of America – lists the following on its website [68]:
1. Metallic pigments – for use in automotive coatings, various electronic materials, and yes, packaging (cans)
2. Chemicals – aluminum in powder form can be part of the synthesis of alcohols and polyolefins
3. Rocket propellants
4. Photovoltaic thick film pastes – solar cell materials
5. Metallurgy – as a chemical reducing agent – aluminum continues to be used in Goldschmidt reactions (established forms of combustions)
6. Refractories – in some applications, in furnace linings
7. Adhesives and sealants
8. Explosives – aluminum powder and iron(III) oxide is used in mining and military explosives.

The use of a mixture of aluminum powder and ferric oxide (aka. iron(III) oxide) has become somewhat better known to the public in the past few years as several of what are called thermite reactions have been posted to social media sites. The reaction contains all the oxygen required for its self-perpetuation, and thus cannot be extinguished by any of the A, B, C, or D fire extinguishers. The U.S. military has for decades issued two thermite grenades with each armored vehicle, in the event that self-destruction is required – if for instance the vehicle is out of fuel and ammunition. One grenade is placed in the gun tube and other on the engine block. Both will burn through to the ground, making the vehicle useless should it be captured by an enemy. Figure 8.12 illustrates the reaction chemistry.

$$Fe_2O_{3(s)} + Al_{(s)} \longrightarrow Al_2O_{3(s)} + Fe_{(l)}$$

Figure 8.12: The thermite reaction.

It is noteworthy that the products shown in Figure 8.12 indicate that iron is a liquid, a molten metal when the reaction is complete.

8.2.6.3 Aluminum recycling

It is vastly less expensive to recycle aluminum metal than it is to refine new metal from bauxite, which is a major driving force for such recycling. For this reason, aluminum recycling programs exist in almost every developed country, although differences in the programs and the laws that promote them differ within provinces and states. While the general public tends to think of aluminum recycling in terms of beverage containers, larger aluminum objects, such as building parts or street lights, are also recycled.

8.2.7 Titanium – Kroll process

The history of titanium is considerably shorter than many elemental metals, as it was discovered and reported first only in 1791. Because of key properties of this element, namely its high strength and hardness, plus its resistance to corrosion even in reactive environments, it has since found numerous uses. Interestingly, the density of titanium is only 4.506 g/cc – higher than aluminum but far lower than iron – which makes it useful either as an elemental metal, or in an array of alloys with other metals of low density, wherever weight becomes an important factor in an application.

While titanium is utilized as an elemental metal, it is one of the metals that sees significant use not only in alloys but as a compound. The alloys are all designed to be used in high temperature or high stress applications. The compound in question is TiO_2.

The two titanium ores from which almost all titanium is reduced are rutile, TiO_2, and ilmenite, generally $FeTiO_3$, although this formula can also be written as $(Fe,Mg, Mn,Ti)O_3$, since these other metal ions can be part of the makeup of an ore batch. Titanium can be produced from these and isolated as ingots, or the reduced metal can be formed as titanium sponge. Figure

The following are large-scale producers of ilmenite:
1. Tellnes Mine, Sokndal, Norway
2. Lac Tinto Mine, in the Rio Tinto Group, Quebec, Canada
3. Richards Bay Minerals, KwaZulu-Natal, South Africa
4. Moma mine, Kenmare Resources, Mozambique
5. Murray Basin, and Eneabba, of Iluka Resources, Australia
6. Indian Rare Earth Mineral Mines, Kerala, India

7. Grande Cote Mine, of TiZir, Ltd., Senegal
8. QIT Madagascar Minerals Mine, in the Rio Tinto Group [87–95]

When ilmenite is reduced to titanium metal, it is often because the metal can be used as an elemental metal that is extremely durable, or for high-performance alloys. Titanium alloys tend to be used because they are extremely hard, and because they are remarkably inert. Titanium is often used in medical applications, such as replacement bone parts, because of its inertness within the human body.

8.2.7.1 Titanium refining and isolation

The Kroll Process is by far the most common means of refining titanium. We will also discuss the older Hunter Process, since the chemistry behind it is still effective, although not as inexpensive.

8.2.7.2 The Kroll process

The reaction chemistry of the Kroll Process is shown in Figure 8.13. Prior to the first reaction, the titanium dioxide starting material must be refined. This removes iron impurities, through what is known as the Becher Process. In the first reaction, the carbon comes from coke.

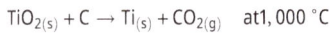

$$TiO_{2(s)} + C \rightarrow Ti_{(s)} + CO_{2(g)} \quad at\,1,000\;°C$$

– which is not isolated, but rather captured as:

$$Ti_{(s)} + Cl_{2(g)} \rightarrow TiCl_{4(g)}$$

Figure 8.13: The Kroll process.

Even though the first reaction makes elemental titanium, it is reacted with chlorine to form titanium(IV) chloride, which is then distilled. The melting point of $TiCl_4$ is −24 °C, and the boiling point is only 136 °C making the distillation a relatively low temperature process. Finally, elemental titanium is isolated. Figure 8.14 shows the reaction.

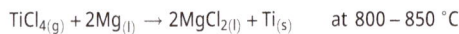

$$TiCl_{4(g)} + 2Mg_{(l)} \rightarrow 2MgCl_{2(l)} + Ti_{(s)} \quad at\;800-850\;°C$$

Figure 8.14: Titanium isolation, the Kroll process.

The end result is titanium sponge. Under an inert atmosphere, the sponge can then be shaped into any form necessary, ingot or otherwise.

8.2.7.3 The Hunter process

The Hunter Process actually produces titanium in a very pure form, and is also the first process by which titanium was refined to the elemental metal. Matthew Hunter's paper concerning this, published over a century ago, in 1910, notes: "All the earliest attempts at the preparation of metallic titanium resulted for the most part in the production of the various nitrides which from their metallic appearance were always mistaken for the metal" [95]. Hunter Process reaction chemistry is shown in Figure 8.15.

$$TiCl_{4(l)} + 4Na_{(l)} \longrightarrow 4NaCl + Ti_{(s)} \qquad at\ 700° - 800\ °C$$

Figure 8.15: The Hunter process.

In an interesting side note, titanium(IV) chloride, $TiCl_4$, has been used in several different organic chemistry reactions. It has also found use in the movie industry as a means by which smokescreens can be formed, since it hydrolyzes readily in air.

8.2.7.4 Uses of titanium, high-strength alloys

The idea that titanium is a high-strength metal is so ingrained in the public mind that for a period of time one of the major credit card issuers even issued a "titanium card." Its high-strength alloys are well known, and have been developed extensively enough that standardized tables for them now exist [98, 99].

We have mentioned titanium when discussing aluminum alloys, simply because depending on compositions, the two can be considered interchangeable. Titanium lends both durability and high tensile strength to alloys made with it, while aluminum creates low density alloys. But many titanium alloys do not require aluminum, and find several other uses. Table 8.12 shows a short list of common titanium alloys.

Table 8.12: Examples of titanium alloys.

ASTM no.	Description	Example use(s)
F136	Ti-6/Al-4/V	Surgical implants
F1295	Ti-6/Al-7/Nb	Surgical implants
B265	Ti alloys (various)	Sheeting and plating
B363	Ti alloys (various)	Welding fittings
B862	Ti alloy	Welded piping

Because so many uses exist for the wide array of titanium alloys, an International Titanium Association advocates for their wide use [99].

In an interesting note, a metal that does not have a large number of uses, ruthenium, is now known to produce high-strength titanium alloys, even with as small an amount as 0.1% ruthenium in the mixture. These alloys can be up to 100 times stronger than titanium in pure form. In the recent past this has developed into a niche use for ruthenium.

8.2.7.5 Titanium recycling

As with every other metal we have examined, titanium is recycled extensively. The market value of the metal is routinely 5–6 times higher than stainless steel. It becomes obvious then that there is a strong economic incentive to recycle elemental titanium and its alloys. The steel industry uses almost all recycled titanium, at times in super-alloys [99].

8.2.8 Platinum group metals or precious metals

So named because of their value, the precious metals, sometimes called the platinum group metals (PGM), are clustered at the lower right of the transition metals, and routinely are considered to include rhodium, iridium, palladium, platinum, silver, and gold.

We have already discussed silver and gold, but will say that since antiquity, silver and gold have been known, and have been highly valued. Ancient religious texts often cite the value of these two metals; and indeed, today they can still be used as a store of wealth. Figure 8.16 shows examples of silver and gold bullion coins which are sold to the public for investment purposes. But silver and gold have also seen extensive use in such areas as photography – when silver emulsions on paper created the image – and in computer contacts – where gold is used.

Beyond silver and gold, the PGMs have relatively recent histories. Platinum was discovered and isolated from South American ores in the 1700s. The remaining were discovered even more recently.

Figure 8.16: Silver and gold bullion coins.

The uses of the PGM are far more diverse than simply a store of wealth, however. For example, both platinum and palladium find wide use in catalytic converters in automobiles. The reaction chemistry is that of the oxidation of carbon monoxide to carbon dioxide, and the reduction of nitrogen compounds to elemental nitrogen.

Production of these two metals, platinum and palladium, is seldom as the primary element or elements from an ore batch. Rather, significant amounts of these metals (and other PGM) comes from the reclamation of what are called anode muds or anode slimes during the production of elemental, high-purity copper. Figure 8.17 shows the basic scheme by which such a reaction occurs.

8.2.8.1 PGM sources, refining and isolation

While PGMs are found and isolated on all six inhabited continents, the dominance of South Africa in current production is because of what is called the Merensky Reef, a portion of the Bushveld Igneous Complex, a major mining region within the country. Its reserves of PGM are believed to be larger than those found in other countries, including nations that have traditionally mined them, such as Bolivia and Russia.

Table 8.13 shows the major refiners of platinum. Besides the PGM, most metals are tracked in tons or thousands of tons. Because of their value and scarcity, PGM tend to be tracked in units of troy ounces, including platinum [9].

Table 8.13: Major producers of platinum [102–112].

No.	Company name	Nation	Amt. (oz)	Other metals
1	Anglo Platinum	South Africa, Zimbabwe	2,378,600	
2	Impala Platinum	South Africa, Zimbabwe	1,582,000	
3	Lonmin Plc	South Africa	687,372	Copper, nickel, iridium, palladium, rhodium, ruthenium
4	Norilsk Nickel	Russia	683,000	Nickel, palladium
5	Aquarius Platinum, Ltd.	South Africa, Zimbabwe	418,461	Palladium
6	Northam Platinum, Ltd.	South Africa	175,000	Gold, iridium, palladium, rhodium, ruthenium, silver
7	Stillwater Mining Co.	USA	154,000	Palladium, rhodium
8	Vale SA	Canada	134,000	Gold, iron, nickel, palladium, silver
9	Xstrata	South Africa, Canada	80,199	Nickel, palladium
10	Asahi Holdings, Inc.	Japan	75,000	

Table 8.14 lists the major producers of palladium, much as Table 8.13 does for platinum. For decades, palladium was valued less than platinum, despite both of them being hard, inert metals that were well able to resist corrosion. In the past five years however, the price of platinum has fallen below that of palladium. This appears to be connected with the increased use of palladium in powder form being used in automotive catalytic converters. Note the similarities in Tables 8.13 and 8.14 in terms of producers. This is a good indicator of how often the two metals are co-located.

Table 8.14: Major palladium producers worldwide.

No.	Name	Location	Amt. (M oz)	Other products
1	Norilsk Nickel	Russia	2.731	Nickel, PGMs
2	Anglo Platinum	South Africa, Zimbabwe	1.395	Platinum
3	Impala Platinum	South Africa, Zimbabwe	1.020	Platinum
4	Lonmin, Plc.	South Africa	0.331	Copper, nickel, iridium, palladium, rhodium, ruthenium
5	Stillwater Mining	USA	0.258	Platinum, rhodium
6	Vale SA	Canada	0.251	Gold, iron, nickel, silver, platinum
7	Aquarius Platinum, Ltd.	South Africa, Zimbabwe	0.238	Platinum
8	North American Palladium	Canada	0.163	
9	Northam Platinum, Ltd.	South Africa	0.096	Gold, iridium, platinum, rhodium, ruthenium, silver
10	Xstrata	South Africa, Canada	0.043	Nickel, platinum

The remaining four PGMs have never been considered stores of wealth as gold and silver have, and have never had major uses. Thus, despite them being hard metals that are excellent in resisting corrosion, they tend to be tracked as a single commodity inclusive of all six elements – PGMs [9].

It is difficult to show how the PGM are separated, because it seldom occurs in a stoichiometrically balanced way, and because all six do not routinely occur in each batch of ore. It is noteworthy though that the separations depend on acidic and basic solutions, and the differing solubilities of the individual metals in each. Also, the starting materials may be anode muds from other refining operations.

in aqua regia: $PGM_{(s)} \rightarrow (Os, Ru, Rh, Ir, Pt)_{(s)} + (Pd, Au)_{(aq)}$

$NaHSO_{4(l)} + (Os, Ru, Rh, Ir)_{(s)} \rightarrow Rh_2(SO_4)_3 + (Os, Ru, Ir)_{(s)}$

$Na_2O_{2(l)} + (Os, Ru, Ir)_{(s)} \rightarrow Ir_{(ppt)} + (Os, Ru)_{(sol)}$

$(Os, Ru)_{(sol)} + O_2 \rightarrow OsO_4 + RuO_4$

$NH_4Cl + OsO_4/RuO_4 \rightarrow (NH_4)_3RuCl_{6(s)} + OsO_4$

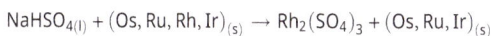

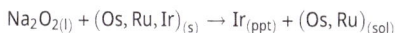

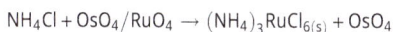

Figure 8.17: General separation and isolation of the platinum group metals.

8.2.8.2 Overall PGM uses

Many of the uses for the PGMs are small or niche ones. A nonexhaustive list includes the following, most of which take advantage of the hardness and nonreactivity of a specific metal. The exception to this is the use of platinum and palladium in automotive catalysts, which takes advantage of these metals selective reactivity:

1. Catalysts in automobiles
2. Chemical processes catalysts
3. (Platinum) crucibles, for laboratory-sized reaction vessels
4. Hard disks for computers (predominantly platinum and ruthenium)
5. Ceramic capacitors
6. Hybridized integrated circuits
7. Dental fillings, dental restoratives
8. Luxury jewelry
9. Various nations' precious metal bullion coinage

8.2.8.3 Uses of ruthenium

Ruthenium has no major, industrial uses, although its hardness can produce alloys in which only a small percentage of ruthenium makes the resulting alloy much harder than the metal which is the major component. This has proven true for titanium alloys especially.

8.2.8.4 Uses of osmium

Osmium is recovered from nickel refining operations and on occasion from copper refining operations. It is such a rare metal that there is very little chemistry devoted to it directly. Iridium-osmium alloys can be extremely hard, and find niche uses in such applications as pen nibs and phonograph needles.

The scientific community has some familiarity with one specific osmium compound, osmium tetroxide – a compound that was traditionally made by direct contact of oxygen gas with powdered osmium metal at high temperature. It has found use in organic synthesis as a way to deliver oxygen cis to a double bond. Its volatility makes it useful in this regard; but it is now often considered prohibitively expensive for such uses.

8.2.8.5 Rhodium

Much like osmium, rhodium is recovered from the anode muds of other refining operations, such as nickel. It is another metal that is scarce enough that very little chemistry involves it as a material. Much more often it is alloyed with platinum or palladium, since the resulting alloys are both hard and corrosion resistant. All objects made from such alloys are usually small, and serve some niche in a specialized industry.

8.2.8.6 Iridium

Very little iridium is produced annually, often less than 10 tons. Like gold and other precious metals, it is more often measured in troy ounces. Because so little iridium exists and is refined, uses for it are necessarily small. One use in the jewelry industry is to coat the prongs of diamond rings to make the gems appear whiter.

While iridium's use as a material is very small, one iridium compound is used extensively, $Ir(CO)_2Cl_2$. It is used catalytically in what is called the Cativa process, which is a means of producing acetic acid from a methanol staring material. It is shown in a simple form in Figure 8.18

8.2.8.7 Palladium

Palladium metal has found a limited amount of use as a store of wealth in the form of palladium bullion coins, issued by a limited number of the world's mints, including those of the United States and Canada. More commonly, palladium in a powdered form is used in catalytic converters in various automobile exhaust systems. Additionally, palladium alloys are used in producing some precious metal jewelry.

8.2.8.8 Platinum metal

As mentioned in our section on gold, platinum can be alloyed with gold, producing what is often called "white gold." This is often 18 parts gold and 6 parts platinum, but other alloys can be made as well. Such alloys can be made into extremely durable jewelry that keeps its look over the course of time.

Platinum has also been minted into bullion coins, in units of one ounce, or of fractions of an ounce. This is a relatively modern means by which a person can own this metal as a store of wealth. Table 8.15 is a listing of nations that produce platinum bullion coins.

8.2.8.9 PGM catalysts

Catalytic converters have utilized platinum as far back as 2008. The key in any such use is to maximize the surface area of the precious metal, to make the use more efficient, and to keep the cost per unit down. The metal surface is capable of reducing nitrogen oxide exhaust gases, ultimately making the mixture which exits at the tail

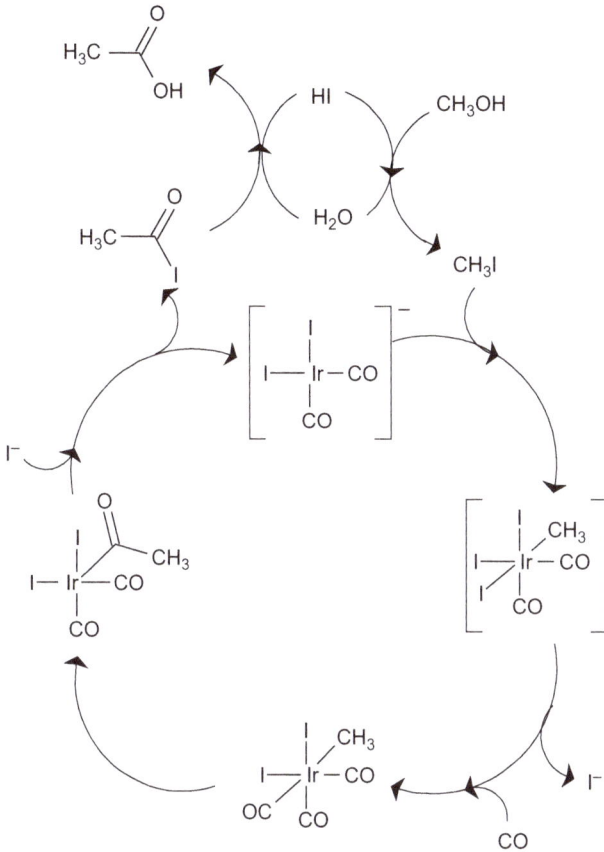

Figure 8.18: Acetic acid production via the Cativa process.

pipe more environmentally friendly. Recent generations of platinum catalysts used in automobiles also use small amounts of rhodium as well.

8.2.8.10 PGM bullion coins

Beyond gold and silver, a select few nations now produce platinum bullion coins as a potential store of wealth for individuals. In the past decade, a few national mints such as the United States Mint and the Royal Canadian Mint have ventured into the production of palladium bullion coins. And in what some metal investors consider something of a publicity ploy, Baird & Company, a British firm, has produced a $100 rhodium coin for the tiny nation of Tuvalu. Table 8.15 lists them.

Table 8.15: Bullion coins, platinum.

Nation	Coin name	Weight
Australia	Koala	1/20, 1/10, ¼, ½, 1 t oz.
Australia	Platypus	1 t oz.
Canada	Maple leaf	1/20, 1/10, ¼, ½, 1 t oz.
Isle of Man	Noble	1/20, 1/10, ¼, ½, 1 t oz.
Mexico	Libertad	¼ t oz.
United States	Eagle	1/10, ¼, ½, 1 t oz.

8.2.8.11 PGM recycling

Because of their high value and their scarcity, all the PGMs are recycled at or near where they are used. When automobiles are scrapped and recycled, the platinum and palladium used in their catalytic converters are also recycled.

8.3 High-performance element and alloys

Several elements that have been discovered more recently than the ancient metals have been found to have excellent properties in terms of hardness, density, or the ability to make alloys with desired properties. While several metals can stake some claim to such properties, we will examine four.

8.3.1 Titanium

As mentioned above, titanium has a relatively short history, having first been discovered in 1791, by William Gregor, and has since found wide use in applications that require low density, high strength, and the ability to withstand heat (its m.p. is: 1,650 °C). Titanium alloys are routinely produced because of their extreme toughness and high tensile strength.

In the aerospace industry, titanium alloys are used in airframes, as well as in missile components where light weight is important, but so is toughness. It is also used in medical implants, such as joint replacements, where interaction with surrounding material should be minimal. Most titanium alloys can be as strong as steel, but weight approximately half of what a steel part would weigh.

8.3.2 Tungsten

Tungsten is not widely distributed throughout the Earth's crust. Rather, it is localized in several areas. Currently, the People's Republic of China produces by far the largest percentage of tungsten, followed rather distantly by Russia [9]. Curiously, even though Portugal produces a relatively small amount of the metal, it was a source that was highly coveted by both sides during the Second World War. Even at that time it was known that tungsten produced alloys that were extremely robust, and that could be put to numerous uses – most of which figured into the war effort of either side.

What are called superalloys are sometimes made with tungsten – with nickel, or iron, or cobalt as well. Two that have become widely known are Stellite and Hastelloy. These have been used in turbine blades, within the aerospace industry, where wear-resistance is critical. Other uses include:
1. Liners for various burners
2. Aircraft engine parts
3. Seagoing vessels propulsion components
4. Components in furnaces, where the ability to withstand high temperatures is important

8.3.3 Magnesium

A low-density metal, magnesium finds significant uses in several industries, such as the aerospace industry. Because of its reactivity, magnesium is not used as an elemental metal; rather, it is always alloyed.

As with several of the other metals we have discussed, magnesium is an element with a relatively short history. It was first discovered in 1808 by Sir Humphry Davy. Its isolation was not trivial, even though one magnesium compound – hydrated magnesium sulfate or Epsom salt – was known and used for over a century prior to this isolation. Magnesium has, however, gained in importance in a wide number of alloys, primarily because of its low density and ability to become part of lightweight alloys.

8.3.3.1 Magnesium refining and isolation

Even though a wide variety of magnesium-containing ores exist, the chief ones from which the metal is refined are dolomite (formula $CaMg(CO_3)_2$) and magnesite (formula $MgCO_3$). Table 8.16 shows a list of the major magnesium-containing types of ore, even though not all of them are economically feasible to use as a source for the metal.

The Pidgeon process is the main means by which magnesium is reduced and refined. It requires some silicon-containing alloy to act as the reducing agent, or a carbon source to serve this purpose. Figure 8.19 shows the process in terms of reaction chemistry, in simplified terms.

Table 8.16: Magnesium-containing minerals.

Name	Chemical formula	Geographic location
Brucite	$Mg(OH)_2$	Pennsylvania, USA
Carnallite	$KMgCl_3 \cdot 6(H_2O)$.	USA, Germany, Russia, Canada
Dolomite	$CaMg(CO_3)_2$	China, Australia, Congo, Morocco, USA, Canada, Brazil, Mexico
Magnesite	$MgCO_3$	South Australia, Brazil, China, India, Russia, Turkey, Spain, Greece, Slovakia, Austria
Olivine	$(Mg,Fe)_2SiO_4$	USA, China, Brazil, Australia, Kenya, Egypt, Mexico, Norway, South Africa, Pakistan, Sir Lanka, Tanzania
Talc	$Mg_3Si_4O_{10}(OH)_2$	Australia, France, Brazil, Oman, Turkey

$$Si_{(s)} + 2CaO_{(s)} + 2MgO_{(s)} \rightarrow 2Mg_{(g)} + Ca_2SiO_4$$

Or:

$$C_{(s)} + MgO_{(s)} \rightarrow Mg_{(g)}CO_{(g)}$$

Figure 8.19: Steps of production of magnesium metal.

A few further notes need to be made to clarify how magnesium is extracted in the Pidgeon process:

1. Required temperature for the first reaction is routinely 1,200 °C–1,500 °C.
2. The reaction requires stainless steel retorts.
3. If ferrosilicon alloy is used, the iron basically is used to deliver the silicon.
4. Since the product is in a gas state, it must then be condensed and ultimately solidified.
5. A temperature of approximately 2,300 °C is required when carbon is the reducing agent.
6. As before, since the product is a gas, it must be solidified before it can be used. Routinely it is shaped into ingots.

It is also possible to extract magnesium from seawater, since a significant amount of it exists in our world's oceans. This reduction chemistry is referred to as the Dow Process. It is a reduction from ionic, dissolved magnesium that is electrolytic. Figure 8.20 shows the simplified reaction chemistry.

The first two reactions do not represent any reduction of the magnesium ions, but rather the separation of them from the much more common sodium ions in seawater. In general, there is only about 12% magnesium in seawater when compared to sodium. This is why it is first separated by differences in solubility – as magnesium hydroxide is not particularly soluble – then regenerated as an aqueous chloride salt.

$$Ca(OH)_2 + MgCl_2 \longrightarrow Mg(OH)_{2(s)} + CaCl_{2(aq)}$$

Followed by:

$$Mg(OH)_{2(s)} + 2HCl \longrightarrow 2H_2O + MgCl_{2(aq)}$$

Followed by:

$$MgCl_2 \longrightarrow Mg + Cl_{2(g)}$$

Figure 8.20: The Dow process for magnesium production.

This ensures that the reduction step begins with a pure magnesium reactant, and not a mixed salt. The final step is the electrolysis which produces the reduced metal and gaseous chlorine as a coproduct.

8.3.3.2 Magnesium: uses

In a discussion of magnesium, there are several compounds that are widely used. But here we will limit the discussion to the reduced metal and to the use of magnesium in metal alloys. The USGS Mineral Commodity Summaries lists the following major uses for the metal:

- "Aluminum-base alloys that were used for packaging, transportation, and other applications accounted for 18% of primary magnesium metal consumption; desulfurization of iron and steel, 4%; and all other uses, 14%."

Other uses combine to form the final 10% [9].

8.3.3.3 Elemental magnesium: uses

In the section concerning titanium, we have discussed the Kroll Process and magnesium's role in it. Because magnesium is easily oxidized, it can be utilized for the reduction of other metals besides titanium, and functions as an effective reducing agent.

An older use for magnesium, one that has waned considerably with the development of new methods of photography, is as a light source in traditional flash photography when flash bulbs were needed. This is because of the intense white light given off when magnesium burns in such flashbulbs

8.3.3.4 Magnesium – aluminum alloys: types and uses

The production of magnesium-aluminum alloys is routinely aimed at making lightweight alloys for use in the aviation and aerospace industry. Magnesium has an extremely low density for a metal, 1.738 g/cm^3. By comparison, the density of aluminum is 2.70 g/cm^3, and iron is 7.874 g/cm^3. Table 8.17 is a nonexhaustive list of magnesium alloys, including some of their properties.

Table 8.17: Magnesium alloys and properties.

Alloyed with	Advantages	Disadvantages	Comments
Aluminum	Strength and hardness	Increasing Al = decreasing ductility	Most common alloy class
Calcium	High-temperature performance	Increasing Ca = decreasing malleability	
Cerium	Corrosion resistance	Lower strength	
Copper	Strength	Lower ductility	
Manganese	Corrosion resistance		
Nickel	Strength	Ductility and corrosion resistance	
Neodymium	Strength	Cost	
Silicon	High-temperature performance		
Tin	Ductility, malleability		
Yttrium	High-temperature strength		
Zinc	Corrosion resistance		Second most common alloy class

8.3.3.5 Magnesium castings

Perhaps more than other metals, magnesium parts as well as magnesium alloy parts often are cast. The ability of such material to flow under the conditions at which they are cast often makes this metal and its alloys ideal for such castings. Final shaving or polishing where two molds come together is often required as a last step.

There is an International Magnesium Association which advocates for the use of magnesium, and which states that common uses for magnesium include: aircraft and missile components, automotive parts, bicycles and bicycle parts, as well as laptop computer components and casings [114, 115].

There are enough magnesium alloys that a system is in place to categorize them. The American Society for Testing and Materials (ASTM) code B275 is used to cover many magnesium alloys, although it was first created for use with other nonferrous alloys. Today, the alloys routinely use two letters and two numbers, such as AZ63 is alloyed with aluminum and zinc, in weight percents of 6% aluminum and 3% zinc.

8.3.3.6 Magnesium in the aerospace industry

Magnesium can be considered a twin of aluminum when it comes to uses in the aerospace industry. Both are low density, and the number of alloys made from them includes some that have excellent combinations of light weight and high strength. In the past decade, research has been directed toward magnesium-lithium alloys, again because of the low density of both metals, and the advantage this has in aerospace applications.

8.3.3.7 Magnesium recycling

As with other metals we have discussed that are used on a large scale, magnesium and its alloys are routinely recycled through scrapyards. As with other metals, the economic incentive is the driving force for such recycling, as this is much less expensive than processing from ores. The USGS Mineral Commodity Summaries comments: "about 25,000 tons of secondary magnesium was recovered from old scrap and 75,000 tons were recovered from new scrap" [9].

Magnesium undergoes one other form of recycling, associated with its use in titanium production. The magnesium chloride that is a coproduct of titanium refining can be re-reduced to the metal, in order for further use in the Kroll process.

8.3.4 Aluminum

As mentioned, aluminum was first discovered in 1827, but the process whereby it is isolated today on an industrial scale, the Hall–Heroult process, was not patented until 1887. During the time in between, aluminum was considered an extremely expensive metal because it was so difficult to isolate from bauxite ores.

One of the least dense metals, and the single least dense to be used and alloyed on a large scale, aluminum finds extensive use in the construction of aircraft. The just-mentioned magnesium alloys and their uses are very often those of aluminum and magnesium.

8.4 The discoveries of the 1700s and 1800s

Numerous elemental metals were discovered in the eighteenth and nineteenth centuries, some almost by accident. In Chapter 1 we have shown, in Figure 1.1, the year in which each element was discovered. Note how many were discovered within the last 250 years.

8.4.1 Lanthanides

As mentioned in the first chapter, Figure 1.1 gives the generally accepted dates at which the elements were discovered, and those discoveries posted. The lanthanides – at times still called the rare earth elements (REEs), or the inner transition elements – were rather latecomers in the overall scheme of things, and the reason why is an aspect of their chemistry that continues to make their separation and purification difficult even today: they all have very similar reactivity. The shape and size of the f-orbitals are such that there is little change in size from the left to the right of the row; and there is little difference in ionic radii or electronegativity. The small differences in ionic radii is often exploited in attempting to separate one of these elements from another. Table 8.18 shows the lanthanides and several of their properties.

Table 8.18: Properties of lanthanides.

Name	Symbol	m.p. (°C)	Ox. state	M^{3+} radius (Å)	$E°$ (V^{-1})
Lanthanum	La	920	+3	1.16	−2.38
Cerium	Ce	795	+3	1.14	−2.34
Praseodymium	Pr	931	+3	1.13	−2.35
Neodymium	Nd	1,016	+3	1.11	−2.32
Promethium	Pm	1,042	+3	1.09	−2.29
Samarium	Sm	1,072	+3	1.08	−2.30
Europium	Eu	826	+3	1.07	−1.99
Gadolinium	Gd	1,313	+3	1.05	−2.28
Terbium	Tb	1,356	+3	1.04	−2.31
Dysprosium	Dy	1,412	+3	1.03	−2.29
Holmium	Ho	1,472	+3	1.02	−2.33
Erbium	Er	1,529	+3	1.00	−2.32
Thulium	Tm	1,545	+3	0.99	−2.32
Ytterbium	Yb	824	+3	0.99	−2.22
Lutetium	Lu	1,663	+3	0.98	−2.30

Note that in Table 8.18, the only property that appears to have significant differences from one element to the next is a physical one, not a chemical one, the melting points. Atomic radii and cell potentials are very close, making the chemistry of any one of the lanthanides quite similar to those of its neighbors, as mentioned.

The name "rare earth element" is something of a misnomer, as several of these elements are more common in the Earth's crust than some of those we think of as common, such as lead and tin. This name was attached to this group of elements approximately at the time of their discovery, simply because they always occur comingled with others of the series, and thus each was rare when it was discovered. Basically, there is no such entity as a lutetium mine, for example. Table 8.19 shows the REEs in terms of their estimated abundance in the Earth's crust, along with some nonrare earths, and their abundance, as points of comparison. The nonrare earth elements in Table 8.19 are shown in italics.

Table 8.19: Rare earth elements and their estimated crustal abundance.

Element	Symbol	Atomic no.	Abundance (ppm)
Zinc	Zn	30	75
Cerium	Ce	58	68
Copper	Cu	29	51
Neodymium	Nd	60	33
Lanthanum	La	57	32
Yttrium	Y	39	30
Cobalt	Co	27	21
Scandium	Sc	21	20
Lead	Pb	82	20
Samarium	Sm	62	20
Gadolinium	Gd	64	20
Praseodymium	Pr	59	16
Dysprosium	Dy	66	13
Ytterbium	Yb	70	10
Hafnium	Hf	72	8
Erbium	Er	68	7
Tin	Sn	50	3
Holmium	Ho	67	3
Terbium	Tb	65	3
Europium	Eu	63	2
Lutetium	Lu	71	2
Thulium	Tm	69	0.7
Bromine	Br	35	0.4
Uranium	U	92	0.03

8.4.1.1 Rare earth elements: sources

Mountain Pass Materials operated the Mountain Pass mine, located in California, close to the Nevada border, from 1952 to 2002. During those years, this mine produced virtually all the europium need for the screens of color televisions. More recently, REE production, usually in the form of rare earth oxides, has been dominated by China. This change has been driven by economic factors more than anything else. The overall cost of extraction and concentration of REEs is lower in China than other nations.

8.4.1.2 REE extraction

As mentioned, the chemistry for the extraction of the rare earth elements continues to be a formidable challenge, based on their similar reactivity. An examination and comparison of the ores containing REEs, often monazite, bastnasite, and xenotime, shows the complexity of them in terms of elements. Table 8.20 lists ores and the rare earth elements in them. The formula for each ore gives some idea of how complex these feedstocks are. Even simple ores, such as fluorite – formally CaF_2 – may include significant amounts of one or more rare earth elements depending upon the batch.

Table 8.20: Ores which contain rare earth elements.

Ore	General formula	Geographic location	Comments
Aeschynite	$(Nd,Ce,Ca,Th)(Ti,Nb)_2(O,OH)_6$	China, Inner Mongolia	May contain significant amounts of Ce, Nd, or Y
Allanite (aka. orthite)	$(Ce,Ca,Y,La)_2(Al,Fe^{3+})_3(SiO_4)_3(OH)$	Greenland; Queensland, Australia; New Mexico, USA	Designated allanite-(Ce), allanite-(La) or allanite-(Y)
Apatite	$Ca_5(PO_4)_3(F,Cl,OH)$	Apatity, Russia; Florida, USA; Canada	Sometimes found without REEs
Bastnasite	$(Ce,La,Y)CO_3F$	Sweden, Pakistan	Notable source of Ce
Brockite	$(Ca,Th,Ce)PO_4 \cdot H_2O$	Colorado, USA	
Cerite	$(Ce,Ca,La)_9(Mg,Fe^{3+})(SiO_4)_6(SiO_3OH)(OH)_3$	Vastmanland, Sweden; Mountain Pass, California, USA; Kola, Russia	Cerite-(Ce) and cerite-(La)
Fluocerite	$(La,Ce)F_3$	Sweden; Kazakhstan; Australia: Inner Mongolia, China	Fluocerite-(La) and fluocerite-(Ce)
Fluorite	CaF_2	Widespread globally	Trace REEs Y, Yb, Eu in fluorite can account for fluorescence
Gadolinite aka. ytterbite	$(La,Ce,Nd,Y)_2FeBe_2Si_2O_{10}$	Norway; Sweden; Colorado, USA	Gadolinite-(Y) or gadolinite-(Ce)
Monazite	$(La,Ce,Pr,Nd,Y,Th)PO_4$	India; Madagascar; South Africa; Bolivia; Australia	Four different possible types
Parasite	$Ca(La,Ce)_2(CO_3)_3F_2$	Colombia; Greenland	Can contain Nd
Stillwellite	$(Ca,Ce,La)BSiO_5$	Queensland, Australia; Tajikistan; Ontario, Canada	
Titanite	$CaTiSiO_5$	Widespread globally	Fe, Al, Ce, Y, and Th possibly present
Wakefieldite	$(L,Ce,Nd,Y)VO_4$	Canada; Congo	Four possible types, based on specific REE
Xenotime	$(Y,Yb,Dy,Er,Tb,U,Th)PO_4$	Brazil; Norway	
Zircon	$ZrSiO_4$	Australia	May also include traces of Hf, U, Th

8.4.1.3 Lanthanide separation and isolation

Simple reaction chemistry which clearly illustrates the separation and isolation of REEs are difficult to illustrate because there are virtually no occurrences in which all of these elements are present in a single batch of ore. The steps required for separation can however be described procedurally, as follows:

1. Milling

The first step in the process, which is similar to that in many other purifications, is milling the batch of material to a uniform particle size. Large pieces of ore are crushed mechanically so that the surface area of the entire batch is increased significantly. Repeated milling steps can bring ores to grain size approximately equal to that of various grades of sand.

2. Electromagnetic division

Some components of REE ores are magnetic. A conveyor belt system can be used to separate such substances in any ore batch from nonmagnetic fractions. In such a system electromagnets are embedded in the end rollers about which the belt rotates. Then milled material drops from the end of the belt into some receiving hopper, the nonmagnetic fraction simply drops, yet the magnetic fraction of the ore is attracted to the embedded magnets within the rotary wheel. This magnetic material within the mixture only drops away from the belt as it pulls away from the embedded magnets, somewhat farther from the end of the wheel.

3. Flotation

The concept of flotation is one of separation of many ores dependent on the density of components and materials within the batch of ore. In all cases, the ore must be brought to powder to increase and maximize the surface area, then be made part of some aqueous solution. Additionally, some type of surfactant material is required in solution. Further, air is injected into such solutions, which can force separate the finer particles those which do not sink.

4. Concentration via centrifuge

The treated and milled ore can be further separated, through the use of high-speed centrifuges, isolating higher-density and lower-density components within any batch of material.

5. Leaching and precipitation

At this point, the concentrated material can usually be separated often as REE oxides, at times as metals, through use of the following:

a. Fractional crystallization

   Differences in solubility among the REEs mean that precipitation can occur in solution. Adjusting temperature and pH under controlled conditions can induce selective precipitation.

b. Ion exchange

This technique concentrates rare earth element ions onto a medium. The medium can be a synthetic plastic resin or some synthetic zeolite. Those ions not attached can be solvated into solution. The materials containing REEs are then washed and further treated with acid, thus isolating the desired products from further ionic materials which remain in solution.

c. Extraction via solvents

Organic solutions, such as kerosene, with a chelating agent in it, can be utilized to extract lanthanides as soluble salts, often as nitrates. Tri-n-butylphosphate is an extensively used chelating material. An agent of this type forms stronger complexes with the lanthanides that have smaller ionic radii, and enhances their solubility in an organic phase.

6. Electrolysis

When metal salts are soluble, an electrolytic deposition of metals can occur. As well, an anodic material, if soluble, can be used during the reduction of a particular REE.

These six broad steps simplify what is normally a very complex series of reaction steps. In some cases, a step needs to be repeated many times (sometimes hundreds of times) to sufficiently enrich a product to a specific REE, and to deplete the others from it.

8.4.1.4 Lanthanide uses

Uses of lanthanides are varied, and continue to expand. One of the most widely known uses is currently the production of high-strength magnets – such as neodymium–iron–boron magnets – that are used in cell phones. First developed in 1982 by General Motors and Sumitomo Special Metals, the aim was not specifically for cellular phone use, but rather to find any possible substitute for the existing samarium–cobalt magnets, which were expensive. Table 8.21 shows some of the more common uses for each of the lanthanides, some of them being quite small when compared to the applications we have seen in this chapter for other metals.

8.4.1.5 Lanthanide recycling

Recycling of any lanthanide-containing user end products remains limited, although it is a growing field. The United States has not yet formulated a national policy for the recycling of such elements. Japan appears to be a world leader in recycling lanthanides, despite the difficulties of re-concentrating them from whatever consumer product they were used in. To provide one example of a use of lanthanides for which recycling can prove worthwhile, and that is practiced in some communities in the United States, cell phones can in some places be recycled in mall kiosks or drop boxes. Figure 8.21 shows one example.

Table 8.21: REEs and their major uses.

Name	Use	Form
Scandium	Aerospace metal alloys	Sc–Al alloy
Yttrium	Ceramics, phosphors, alloys	$EuYVO_4$ for phosphors
Lanthanum	Batteries, catalysts for petroleum refining	NiMH battery component
	Glass	As LaF_3
Cerium	Catalysts, auto	Ce_2O_3
	Oven cleaning catalyst	Ce_2O_3
	Optical polishing	CeO_2
	Ceramics	CeO_2
	Optician's rouge	CeO_2
	Gas mantles	CeO_2
	Alloys	Ce metal
Praseodymium	Magnet alloys	Pr–Nd metal
	Carbon arc lights	$(La)F_x$
	Glass	Mixed
	Fiber optic amplifier	Mixed
	Mischmetal	Pr, Ce, La, Nd
Neodymium	Magnets	NdFeB alloy
Promethium	Beta radiation emitter	As oxide or chloride
Samarium	Magnets	Sm–Co alloy
	Reactor control rods	Alloy component
Europium	Liquid crystal displays	Eu_2O_3
Gadolinium	Numerous niche uses	Usually as alloy component, or as oxide
Terbium	Phosphors	$Tb:Gd_2O_2S$
	Magnets	Minor component
Dysprosium	Magnets	Nd–Fe–B–Dy, up to 5% Dy
	Lasers	Dy–V
Holmium	High-power magnets	As alloy
Erbium	Lasers	$Y_3Al_5O_{12}$:Er
	Specialty glasses	As a dopant
Thulium	X-ray device source	As reduced metal
	Lasers	$Tm:Ho:Cr:Y_3Al_5O_{12}$
Ytterbium	Stainless steel	Minor dopant
	Fiber optics	As dopant
Lutetium		Few commercial uses

Like many other recycling programs, it is an economic incentive that drives the program, and not one related to the scarcity of any of the materials. While it is difficult to predict how large the reserves of lanthanide-containing ores are, some experts believe that more than a century of rare earth elements remain left that can be mined, refined, and extracted from ores. Recycling may still prove to be more cost effective in the long run.

Figure 8.21: Cellular phone recycling kiosk in a mall.

8.5 Ferrous-based alloys

We have already mentioned steel, the class of alloys of iron that finds the greatest number of uses. But the term "steel" now incorporates iron mixed with up to 4% carbon, yet often with small amounts of some other metal or metals as well. Table 8.22 is a nonexhaustive list of steels – as a full list would be pages long – gives an example ASTM number (since most have more than one number for a class of steels), and gives examples of some of their uses.

The reason so many different steel formulations exist is because of the enormous number of niche uses that a particular steel might have. For example, the term "stainless steel" implies an alloy that is corrosion resistant, but some are designed specifically for use in a saline marine environment – the oceans – while others are designed for piping use in highly acidic environments – in chemical plants, for example.

8.6 Other metal alloys

There are a truly huge number of metal alloys that do not involve iron or the other major metals. We discuss only a select few of them below.

Name	Example ASTM no.	Composition	Example use(s)	Comments
Austenitic stainless steel	ASTM A269		Tubing and piping	
Carbon steel	ASTM A283			Four different grades
Chrome steel	ASTM A387		Pressure vessels	
Chromium–vanadium steel	ASTM A231	C-0.50%, Cr-0.95%, V-0.15%	Spring wire	
Ferritic stainless steel	ASTM A493	Cr-11.0%, Si-1.0%, Mn-1.0%, Ni-0.5%	Excellent ductility and corrosion resistance	Multiple grades and ASTM numbers
Martensitic stainless steel	ASTM A217		High-temperature applications	

8.6.1 Low melting alloys

Low melting alloys are another class of materials that find a wide array of uses in modern society. The term "low melting" has different meanings depending upon the context in which it is used.

For an example of a low melting alloy that involves a liquid metal at room temperature, sodium–potassium alloys (sometimes simply called NaK alloys, and pronounced "nack") are used in small amounts in dry boxes (aka. glove boxes) in scientific laboratories. They can exist as a liquid at room temperature, and the liquid metal surface scavenges any residual oxygen that is in a nitrogen-filled or argon-filled glovebox, keeping it at a high state of purity.

Another such example of a liquid, low-melting alloy is galinstan – so named because it is made from gallium, indium, and tin. This alloy remains a liquid to $-19\,°C$, and thus finds use in some specialty, low temperature thermometers.

For an example of a low melting alloy that is solid at room or temperature, but that can melt in boiling water, Wood's metal fusible alloy is a well-known material (also known as Lipowitz's alloy), having been first discovered in the mid-1800s. The alloy has a composition of: bismuth 50%, lead 26.7%, tin 13.3%, and cadmium 10% by weight, and melts at 70 °C. Wood's metal was first produced by Dr. Barnabas Wood in 1860, but continues to have several important uses today. It is used as a solder, as a means of holding parts in place, and perhaps most ubiquitously, as a valve plug in fire sprinkler systems in buildings.

Figure 8.22 shows a sample of Wood's metal ingots. Figure 8.23 shows a sample of Wood's metal melting in a beaker of water on a laboratory hot plate.

Figure 8.22: Wood's metal fusible alloy ingots.

Figure 8.23: Wood's metal fusible alloy melting.

There are certainly other low melting alloys. Table 8.23 is a list of them, but is certainly not inclusive of all.

Table 8.23: Low melting alloys.

Alloy	m.p. (°C)	Composition (wt%)	Example uses
Bi-Pb-Sn-Cd-In-Tl	41.5		
Cerrolow 136	58	Bi-Pb-Sn-In	
Field's metal	62	Bi32.5-Sn16.5-In51	
Galinstan	−19	Sn 10.0, In 22, Ga 68	Low-temperature thermometers
Onion's metal	92	Bi 50, Pb 30, Sn 20	Cadmium-free alloy
Rose's metal	98	Bi 50, Pb 25, Sn 25	
Wood's metal	70	Bi 50, Pb 26.7, Sn 13.3, Cd 10	

8.6.2 Lanthanide-containing alloys and others

There are several alloys that contain one or more of the lanthanide elements, often with the lanthanide being only a small portion of the material by mass. Two rather noteworthy examples are discussed here.

8.6.2.1 Super magnets

What is now called a neodymium–iron–boron magnet, or Nd–Fe–B magnet, was first patented in 1982, as mentioned, and was a collaboration between General Motors and Sumitomo Metals. Sometimes called NdFeB, or "Neo," or "NIB," or super magnets, the amount of neodymium in such magnets varies, but iron remains the main component. The formula $Nd_2Fe_{14}B$ is one that has been mass-produced for a variety of applications in which small magnets can be used productively.

Curiously, the strength of these Nd–Fe–B magnets is such that at least one company sells them in pairs, often to science teachers, with a piece of wood between the two. This also comes with a written warning that if the wood is removed and the magnets stuck together by the user, the company will not provide a refund for any return!

8.6.2.2 Tantalum–niobium wire

The metals tantalum and niobium have not been discussed elsewhere in this chapter because both are rare elements, and their uses are somewhat limited. One important use, however, is the production of tantalum-niobium wire which becomes the coil for an NMR or MRI instrument. As we might expect, there are relatively few companies that have the ability to produce such wire, especially in the lengths needed for a magnetic coil in either an NMR or an MRI. All companies choose to keep their compositional formulas proprietary.

8.7 Novel metal architectures (e.g., foams)

The characteristics and properties of metals are often determined by their strength, their ability to conduct electricity, their magnetic properties, or their density. The density of metals is often a positive attribute of how they behave under stress conditions, but can also be a detraction when light materials are required. Thus, metal foams are sometimes used when weight is a critical factor.

Metal foams are materials that are composed of various metals, but the metal is positioned much like the solid vertices of a sponge, with air or some other gas, usually an inert one, in the open pores. The analogy to a sponge is somewhat limited however, in that the metal portion of the foam is rigid and unmoving. Still, no more than 25% of a metal foam is metal, and the pores can be connected – an open cell foam – or isolated – a closed cell foam. Since light weight is the reason metal foams are produced, most utilize aluminum as the metal.

The production of metal foams can be considered a junction of art and science, since the air or other foaming gas must be injected into the molten material at the proper temperature. The metal should be molten, but should also be viscous enough that bubbles are trapped in it when it cools, and do not all simply escape from the molten metal.

Another means by which metal foams can be produced is to introduce a liquid or second solid into the solid metal, then the entire mixture heated to the point where the liquid is gasified. The gas is thus released into the melt, and trapped as bubbles in it when the entire material re-solidifies.

Applications for metal foams tend to be niche markets, since most metal foams are expensive to produce, certainly when compared to traditional metals. They have found use as heat exchangers, and as shock absorbers. In the first case, the bubbles tend to slow the transmission of heat. In the second, the porous nature of the material absorbs mechanical shocks.

8.8 Magnetic materials

Magnetism – what is more properly termed ferromagentism – is the arrangement of atoms in a sample of iron in which the bulk material attracts other iron-based materials, or is attracted to them. Naturally occurring magnets have been known since antiquity, and still carry the name lodestone. A lodestone is a piece of magnetite that naturally possesses iron atoms in an arrangement that produces a magnetic field.

Iron remains the base metal used for most magnets, despite extensive research into metals, alloys, single molecules and other such systems that are nonferrous. Cobalt, nickel, gadolinium, and at low temperatures dysprosium can also display magnetic behavior. But other elements sometimes are incorporated and are found to be useful in such formulations as well. It is noteworthy that almost all of these elements, as well as iron, are near the center of the d-block or f-block of the Periodic Table of the elements, and thus naturally have a high number of unpaired electrons. Examples of such magnets are listed below. These are permanent magnets, meaning they retain their magnetic properties even when not in the presence of some external magnetic field (a field from some other source).

- Alnico: Several alloys have the name alnico. This is a permanent magnet still including iron, but that includes aluminum, nickel, and cobalt in their compositions. Copper can also be included in such alloys, and more recently, titanium. Such magnets can be made by casting, in which the components are liquefied at high temperature and mixed. They can also be produced by sintering, a process in which both heat and pressure are applied, but during which the components are not melted.
- Ferrite: This is an iron(III) oxide (Fe_2O_3) powder blended in a ceramic base material, which may also have in it small amounts of such elements as nickel, zinc, or manganese.
- Samarium–cobalt: First developed in the 1960s, different formulations have been developed, with $SmCo_5$ now being one of the most common. They have proven to be highly resistant to demagnetization in a variety of environments.
- Neodymium–iron–boron: Sometimes called "NIB" magnets, these were developed in the 1980s, when the formula $Nd_2Fe_{14}B$ was found to be a highly effective permanent magnet. The use of such magnets in cellular phones, and the sharp increase in the number of such devices in the past two decades, has put stress on the sourcing of new deposits of ores which contain neodymium.

Figure 8.24: A ferrofluid attracted to a magnet.

Related to magnetic materials are what are called ferrofluids. As the name implies, these are liquids that respond to a magnet. First developed by NASA to solve the problem of how to direct some liquid in a micro-gravity environment, ferrofluids have found other niche uses as well. Figure 8.24 shows a simple ferrofluid made in a teaching laboratory. The bar magnet attracting the particles in the fluid can be seen at the bottom of the photo.

8.9 ASTM standards for metals

The enormous number of metal alloys that have found uses in specific applications requires some form of standardization. The ASTM has become an organization that monitors such standards. Databases of numeric designators now exist, and are extensive. Such a listing is beyond the scope of this book, but such can easily be found online [115].

8.10 Recycling

As seen throughout this chapter, the recycling of metals is both a large industry – such as the recycling of iron and aluminum – and yet one that is still evolving – such as the recycling of lanthanides. The precious metals are routinely recycled even if they have been used in small amounts, simply because of their value.

The challenges to this appear at the present time to determine how to recycle relatively rare metals that are widely dispersed, and exist in small amounts in an enormous number of consumer end use products. The cell phones that have been mentioned in the discussion of lanthanides are a prime example. A single neodymium–iron–boron magnet does not use much neodymium. Yet if such magnets are simply discarded, the supply of neodymium will decrease over time. Clearly, intelligent strategies concerning how to recycle items such as cell phones and other products that use lanthanides will have to be developed in the future.

References

[1]   American Society for Testing and Materials, ASTM. Website. (Accessed 14 June 2025, as: https://www.astm.org).

[2]   M. Friedman. An archaeological dig reignites the debate over the old testament's historical accuracy, *Smithsonian*, December 2021.

[3]   G. Zhao, W. Zhang, Z. Duan, M. Lian, N. Hou, Y. Li, S. Zhu, S. Svanberg. Mercury as a geophysical tracer gas – emissions from the emperor Qin Tomb in Xi'an studief bylaser radar, *Nature*, nature.com/articles/s41598-020-67305-x.

[4]   Vale. Website. (accessed 14 June 2025, as: https://vale.com).

[5]   Rio Tinto. Website. (accessed 14 June 2025, as: https://www.riotinto.com).

[6]   BHP. Website. (accessed 14 June 2025, as: https://www.bhp.com).

[7]   Fortescue Metals Group. Website. (accessed 14 June 2025, as: https://www.fortescue.com).

[8]   Anglo American. Website. (accessed 14 June 2025, as: https://www.angloamerican.com).

[9]   USGS Mineral Commodity Summaries 2024, downloadable.

[10] American Iron and Steel Institute. Website. (Accessed 14 June 2025, as: https://www.steel.org)

[11] Eurofer, the European Steel Association. Website. (Accessed 14 June 2025, as: https://www.euro fer.eu).

[12] Canadian Steel Producers Association. Website. (Accessed 14 June 2025, as: https://canadian steel.ca).

[13] Australian Steel Association. Website. (Accessed 14 June 2025, as: https://www.steelassociation. com.au).

[14] The Japan Iron and Steel Federation. Website. (Accessed 14 June 2025, as: https://www.jisf.or.jp).

[15] Korea Iron and Steel Association. Website. (Accessed 14 June 2025, as: https://www.kosa.or.kr).

[16] ArcelorMittal. Website. (Accessed 14 June 2025, as: https://corporate.arcelormittal.com/).

[17] Posco. Website. (Accessed 14 June 2025, as: https://www.poscointl.com/eng).

[18] Nippon Steel Corporation. Website. (14 June 2025, as: https://www.nipponsteel.com).

[19] JFE Holdings. Website. (Accessed 14 June 2025, as: https://www.jfe-holdings.co.jp/en/).

[20] Baosteel. Website. (Accessed 14 June 2025, as:https://www.baosteel.com).

[21] Tata Steel. Website. (Accessed 14 June 2025, as: https://www.tatasteel.com/).

[22] Nucor. Website. (Accessed 14 June 2025, as: https://nucor.com/).

[23] Metalurgica Gerdau. Website. (Accessed 14 June 2025, as: https://www2.gerdau.com).

[24] Kobe Steel, Ltd. Website. (Accessed 14 June 2025, as: https://www.kobelco.co.jp/english/).

[25] World Steel Association. Website. (Accessed 14 June 2025, as: https://worldsteel.org/).

[26] World Steel Organization. Website. (Accessed 14 June 2025, as: https://worldsteel.org/data/world-steel-in-figures/world-steel-in-figures-2024/).

[27] The Iron and Steel Society: A Division of the Institute of Materials, Minerals, and Mining. Website. (Accessed 14 June 2025, as: https://www.issource.org).

[28] The Iron and Steel Institute of Japan. Website. (Accessed 14 June 2025, as: https://www.isij.or.jp/en glish).

[29] European Confederation of Iron and Steel Industries. Website. (Accessed 14 June 2025, as: https://www.eurofer.eu).

[30] African Iron and Steel Association. Website. (Accessed 14 June 2025, as: https://www.uia.org/).

[31] South African Iron and Steel Institute. Website. (Accessed 14 June 2025, as: https://www.saisi.org).

[32] Australian Steel Institute. Website. (Accessed 14 June 2025 as: https://www.steel.org.au/).

[33] British Stainless Steel Association. Website. (Accessed 14 June 2025 as: https://bssa.org.uk/).

[34] International Nickel Study Group. Website. (Accessed Chapter 14 June 2025 as: http://insg.org/).

[35] International Stainless Steel Forum. Website. (Accessed 14 June 2025 as: https://www.ssina.com).

[36] Steel Recycling Institute. Website. (Accessed 14 June 2025 as: http://gssd.mit.edu/search-grad/site/steel-recycling-institute).

[37] American Iron and Steel Institute. American Steel Sustainability. Website. (Accessed 29 March 2022, as: https://steel.org/sustainability).

[38] European Steel Association (EUROFER). Environment – Eurofer. Environmental protection is of top importance for the steel industry. Website. Accessed 14 June 2025 as: https://www.euofer.eu/is sues/environment).

[39] National Slag Association. Website. (Accessed 14 June 2025, as: https://nationalslag.org/).

[40] J.D. Verhoeven, A.H. Pendray, W.E. Dauksch, D. Pendray. The key role of impurities in ancient damascus steel blades, *Journal of Metallurgy*, 1998, 50(9): 58. Bibcode:1998JOM . . . .50i.58V. doi: 10.1007/s11837-998-0419-y

[41]   A. Tucker. Going for the gold, *Smithsonian*, July – August 2012, 34–36.
[42]   World Gold Council. Website. (Accessed 14 June 2025, as: https://www.gold.org).
[43]   World Gold Panning Association. Website. (Accessed 14 June 2025, as: https://www.worldgoldpanningassociation.com).
[44]   Gold Prospectors Association. Website. (Accessed 14 June 2025, as: https://www.gpaastore.com).
[45]   Barrick Gold Corp. Website. (Accessed 14 June 2025, as: https://www.barrick.com).
[46]   Newmont Mining Corp. Website. (Accessed 14 June 2025, as: https://www.newmont.com).
[47]   AngloGold Ashanti. Website. (Accessed 14 June 2025, as: https://www.anglogoldashanti.com).
[48]   Gold Fields. Website. (Accessed 14 June 2025, as: https://www.goldfields.com/).
[49]   Newmont Corporation. Website. (Accessed 14 June 2025, as: https://www.newcrest.com).
[50]   Kinross Gold Corporation. Website. (Accessed 14 June 2025, as: https://www.kinross.com).
[51]   U.S. Gold Corp. Website. (Accessed 14 June 2025, as: https://www.usgoldcorp.gold).
[52]   Pan American Silver. Website. (Accessed 14 June 2025, as: http://yamana.com).
[53]   Agnico-Eagle Mines. Website. (Accessed 14 June 2025, as: https://www.agnicoeagle.com).
[54]   Polyus Gold. Website. (Accessed 14 June 2025, as: https://www.polyus.com).
[55]   Codelco. Website. (Accessed 14 June 2025, as: https://www.codelco.com).
[56]   Freeport-McMoRan. Website. (Accessed 14 June 2025, as: https://fcx.com).
[57]   BHP Billiton. Website. (Accessed 14 June 2025, as: https://www.bhp.com).
[58]   Xstrata plc. Website. (Accessed 14 June 2025, as: https://www.glencore.com).
[59]   Rio Tinto Group. Website. (Accessed 14 June 2025, as: https://www.riotinto.com).
[60]   Anglo American Plc. Website. (Accessed 14 June 2025, as: https://www.angloamerican.com).
[61]   Grupo Mexico. Website. (Accessed 14 June 2025, as: https://www.gmexico.com).
[62]   Glencore International SA. Website. (Accessed 14 June 2025, as: https://www.glencore.com).
[63]   Southern Copper Corp. Website. (Accessed 14 June 2025, as: https://southerncoppercorp.com).
[64]   KGHM Polska Miedz. Website. (Accessed 14 June 2025, as: https://kghm.com).
[65]   Copper Development Association, Inc. Website. (Accessed 14 June 2025, as: https://www.copper.org/).
[66]   Bronze.net. Website. (Accessed 14 June 2025, as: http://www.bronze.net).
[67]   UC Rusal. Website. (Accessed 14 June 2025, as: https://rusal.ru/en).
[68]   Alcoa, Inc. Website. (Accessed 14 June 2025, as: https://www.alcoa.com).
[69]   Aluminum Corp. of China. Website. (Accessed 14 June 2025, as: https://www.chalco.com.cn).
[70]   China Power Investment Corp. Website. (Accessed 14 June 2025, as: https://www.cccme.cn).
[71]   Rio Tinto Alcan Inc. Website. (Accessed 14 June 2025, as: https://www.riotinto.com/canada).
[72]   Norsk Hydro. Website. (Accessed 14 June 2025, as: https://www.hydro.com).
[73]   China Hongqiao. Website. (Accessed 14 June 2025, as: http://en.hongqiaochina.com).
[74]   Shandong Weiqiao Aluminum & Power Group Project. Website. (Accessed 14 June 2025, as: https://en.sdgzgf.com/case).
[75]   Shandong Xinfa Aluminum & Electricity Group Ltd. Website. (Accessed 14 June 2025, as: https://xinfaalum.com).
[76]   EGA: Emirates Global Aluminum. Website. (Accessed 14 June 2025, as: https://www.ega.ae/en).
[77]   International Aluminium Institute. Website. (Accessed 14 June 2025, as: https://world-aluminium.org).
[78]   The Aluminum Association. Website. (Accessed 14 June 2025, as: https://www.aluminum.org).
[79]   European Aluminium Association. Website. (Accessed 14 June 2025, as: https://european-aluminium.eu).
[80]   Australian Aluminium Council. Website. (Accessed 14 June 2025, as: https://aluminium.org.au).
[81]   Aluminium Association of Canada, Dialog on Aluminum. Website. (Accessed 14 June 2025, as: https://aluminium.ca).

[82] AFSA – Aluminium Federation of South Africa. Website. (Accessed 14 June 2025, as: http://www.afsa.org.za/).

[83] Aluminium Association of India. Website. (Accessed 14 June 2025, as:https://www.indiamart.com/aluminium-association-india/aboutus.html).

[84] Alcoa. Website. (Accessed 14 June 2025, as: https://www.alcoa.com/foundation/en

[85] ArcticEcon. Website. (Accessed 14 June 2025, as: https://arcticecon.wordpress.com/category/metals-and-minerals/aluminum).

[86] Aluminum.org. "International alloy designations and chemical composition limits for wrought aluminum and wrought aluminum alloys," Website. (Accessed 14 June 2025, as: http://www.aluminum.org/sites/default/files/2021-10/Teal%20Sheet.pdf).

[87] Kronos Worldwide, Inc. Website. (Accessed 14 June 2025, as: https://kronostio2.com).

[88] Canada – Rio Tinto. Website. (Accessed 14 June 2025, as: https://www.riotinto.com).

[89] Richards Bay Minerals – Rio Tinto. Website. (Accessed 14 June 2025, as: https://riotinto.com/en/operations/south-africa/richards-bay-minerals).

[90] Kenmare Resources plc. Website. (Accessed 14 June 2025, as: https://www.kenmareresources.com).

[91] Iluka. Website. (Accessed 14 June 2025, as: https://iluka.com).

[92] Indian Rare Earths Minerals. Website. (Accessed 14 June 2025, as: https://www.irel.co.in).

[93] Eramet. Website. (Accessed 14 June 2025, as: http://www.tizir.co.uk).

[94] QIT Madagascar Minerals. Website. (Accessed 14 June 2025, as: riotinto.com/en/operations/madagascar/qit-madagascar-minerals).

[95] M.A. Metallic Titanium. Hunter, *Journal of the American Chemical Society*, 1910, 32(3): 330–336.

[96] Alcoa. Website. (Accessed 14 June 2025, as: https://www.alcoa.com/global).

[97] Space Materials Database. Website. (Accessed 14 June 2025, as: https://www.spacematdb.com).

[98] ASTM International. Website. (Accessed 14 June 2025, as: https://store.astm.org/b0367-13r17.html).

[99] International Titanium Association. Website. (Accessed 14 June 2025, as: https://titanium.org/).

[100] F. Haber US Patent 1202995. Production of Ammonia. US1202995.

[101] International Platinum Group Metals Association. Website. (Accessed 14 June 2025, as: https://ipa-news.com/).

[102] Anglo Platinum. Website. (Accessed 14 June 2025, as: https://www.angloamerican.com).

[103] Impala Platinum. Website. (Accessed 14 June 2025, as: https://www.implats.co.za).

[104] Sibanye-Stillwater. Website. (Accessed 14 June 2025, as: https://www.sibanyestillwater.com).

[105] Nornickel. Website. (Accessed 14 June 2025, as: https://nornickel.com).

[106] Aquarius Platinum, Ltd. Website. (Accessed 14 June 2025, as: https://www.sibanyestillwater.com).

[107] Northam Platinum, Ltd. Website. (Accessed 14 June 2025, as: https://www.northam.co.za).

[108] Sibanye-Stillwater. Website. (Accessed 14 June 2025, as: https://www.sibanyestillwater.com).

[109] Vale SA. Website. (Accessed 14 June 2025, as: https://vale.com).

[110] Xstrata, Glencore plc. Website. (Accessed 14 June 2025, as: https://www.glencore.com).

[111] Asahi Holdings, Inc. Website. (Accessed 5 June 2022, as: https://www.asahigroup-holdings.com).

[112] Impala Canada. Website. (Accessed 14 June 2025, as: https://www.nap.com).

[113] National Cancer Institute. The "Accidental" Cure – Platinum-based Treatment for Cancer: The Discovery of CisplatinWebsite. (Accessed 14 June 2025, as: http://www.cancer.gov/research/progress/discovery/cisplatin).

[114] International Magnesium Association. Website. (Accessed 14 June 2025, as: https://www.intlmag.org/).

[115] Total Materia. Introduction to the ASTM Designation System. Website. (14 June 2025, as: https://totalmateria.com/en-us/articles/introduction-to-the-astm-system).

Chapter 9
Defect structures

9.1 Introduction

In nature, naturally formed crystals do not exist as perfectly ordered matrices throughout their lattice structures at the atomic or ionic level, although some may come close. Even in a laboratory setting, it is very difficult to grow crystals of a purity level that is acceptable for specific applications, such as for computer chips. In general, the presence of some type of defect is not automatically a problem, as the defects often tailor or adjust the properties of the material. For example, some defects attenuate the conductivity of a material, which we will explore in some detail in Chapter 11.

One extremely common example of a defect that alters the properties of a material has already been discussed in Chapter 8, steel. The addition of small amounts of elemental carbon to iron – in what can be considered an "extrinsic interstitial defect" – creates steel, and makes the resulting material much stronger than pure, elemental iron. To be fair, the many variations on the formula for steel are such that scientists and engineers usually categorize them simply as alloys, and not as a pure material that has some defect or defects within it.

Additionally, natural crystalline materials always have some impurities incorporated within them, which are a form of defect [1–3]. Some of these actually make the end object famous in some way. One well-known and famous crystal is the Hope Diamond. It is blue, and not clear, as the finest diamonds often are. The reason for this blue color is apparently because of minor amounts of boron impurities that formed naturally within it. But many crystalline materials besides diamonds are also graded in terms of what impurities exist within them. Professionals who deal in diamonds often speak of "four Cs," meaning color, cut, clarity, and carat weight. The first and third of these characteristics relate to the possibilities of defects within the diamond.

One example of the need to control the amount of imperfections within crystals, one in which purity is very pertinent to the end use of the material, is the growth of extremely pure silicon crystals. Silicon is always obtained from silica (SiO_2). This elemental silicon is then purified to a high degree, 99.9999%, using what is called a zone refining furnace, as shown in Figure 9.1.

The refining of silicon dioxide (SiO_2) to produce the silicon that is then purified in such a furnace is shown in Figure 9.2. Note that elemental carbon is used as the reducing agent.

The vertical design of a zone refining furnace utilizes gravity to help concentrate any impurities within the silicon crystal. The impurities are concentrated in the heated portion of the silicon crystal, and move with this semi-molten portion of the crystal, always in a downward direction. This process of moving the donut-shaped

https://doi.org/10.1515/9783112205822-009

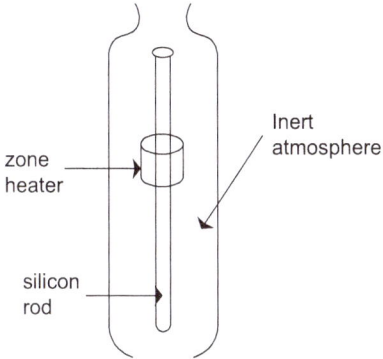

Figure 9.1: Zone refining furnace.

$$SiO_{2(s)} + 2C_{(s)} \rightarrow 2CO_{(g)} + Si_{(s)}$$

Figure 9.2: Silicon production from silica.

furnace down the silicon crystal or rod can be repeated as needed, and ultimately the lowermost portion of the silicon – that containing the impurities – can be cut off.

But this means of producing as pure a crystal of silicon as is possible is not always required in the production of other solids. Even in the production of highly refined silicon, in which the just-mentioned purity of 99.9999% is attained, there are still defects in the finished product.

We will discuss the presence of defects in different types of solids, and look at examples in which the defect in some way alters the macroscopic properties of it.

9.2 Intrinsic point defects

Defects can be categorized in different ways. We will divide them into intrinsic and extrinsic point defects, and into extended defects. An intrinsic point defect is broadly a defect in which an atom is missing – creating a vacancy – or an extra atom is present – often called an interstitial. An extrinsic point defect involves the presence of some added impurity beyond the atoms or ions that are normally in the crystalline material. And as the name implies, extended defects are in more than one position, and can be ordered in two or even three dimensions within a material.

9.2.1 Frenkel defects – aka, interstitial defects

Named after the first person to discover and study them, Yakov Frenkel (also Jakov, 1894–1952), these defects are formed when some particle – an atom, or often a small cation – is displaced from a lattice, leaving a hole or vacancy. The particle does not leave the crystalline structure – it is not ejected – but rather occupies an interstitial space within it. Figure 9.3 shows an idealized example of this. Note both the hole and the particle that appears to be resting atop the layer of cations and anions.

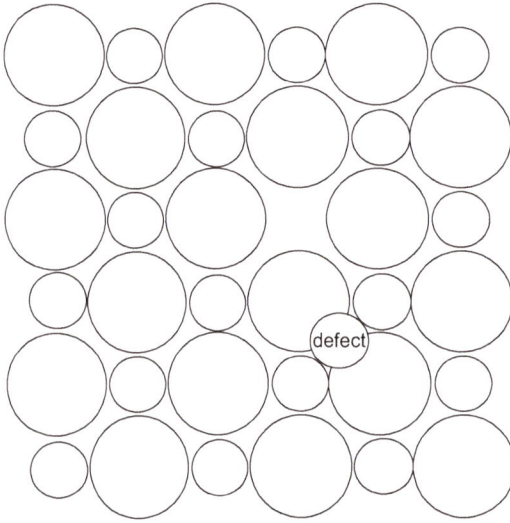

Figure 9.3: Frenkel defect.

Frenkel defects can occur with either cations or anions being displaced. Such displacements do not change the stoichiometric ratio of the particles in the material. Rather, the common factor in Frenkel defects is that generally it is the smaller ion that is more easily displaced into an interstitial site. When cations and anions in a salt crystal are of similar size, it is possible to have Frenkel defects in which both types of ions are displaced into interstitial sites.

9.2.2 Schottky defects – aka, vacancy defects

Named after Walter Hans Schottky (1886–1976), these defects form when an atom or ion moves out from a position somewhere in the interior of a lattice to that crystal's surface [4, 5]. They are sometimes referred to as vacancy defects because of the resulting holes in the lattice structure. To maintain electronic neutrality throughout any material, such defects consist of vacancies in equal amounts in the cation and anion of an ionic crystalline structure. Figure 9.4 shows an example of this.

As with Frenkel defects, Schottky defects do not affect the stoichiometric ratio of the material in which they occur.

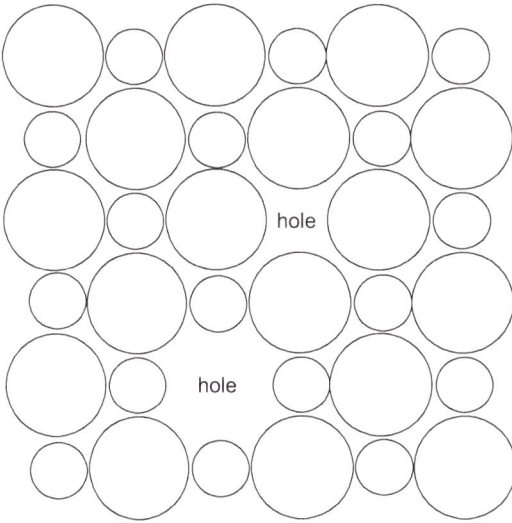

Schottky defect.

Schottky defects tend to occur more in highly ionic crystalline solids than do Frenkel defects, simply because the ionic nature of such a solid means that the loss of matched cation and anion pairs is favored over the dis-placement or re-placement of a single ion in close proximity to others of equal charge. Perhaps the most common example of a material in which Schottky defects can occur is rock salt, sodium chloride, although salts that are not a one-to-one MX ratio can also have Schottky defects. Also, materials which have high coordination numbers, such as metals that exist in close-pack arrays, tend to manifest Schottky defects.

Beyond Frenkel and Schottky defects, materials such as metal alloys can exhibit what are called antisite defects. As the name implies, when an alloy has a regular structure, one in which the atomic radii and the electronegativity of the two elements are close, one atom can take the place of the other. Figure 9.5 gives a stylized example. The dislocation is at the center-left of the figure.

9.3 Extrinsic point defects

An extrinsic point defect is defined as the existence of some atom in the crystal structure that is foreign to it and its repeat pattern. These are sometimes referred to as dopants, and are routinely incorporated into semiconductors, which we will discuss in more detail in Chapter 11. Figure 9.6 gives a basic example. As an example, if the material shown is silicon, and the shaded atom was a gallium atom, the structure would have one less electron in that locus. If the shaded atom were arsenic, the area would have one more electron that the surrounding atoms. Since such dopants occur

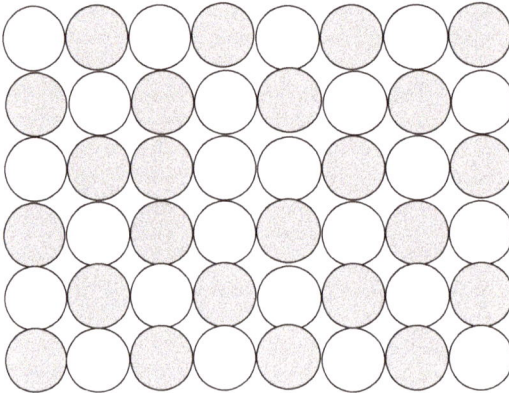

Figure 9.5: Antisite defect.

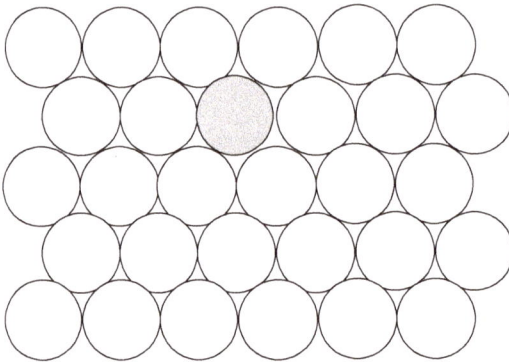

Figure 9.6: Dopant atom in a material.

many times in a material, what is called a conduction band begins to be populated. We will discuss this in more detail in Chapter 11.

Another type of extrinsic point defect is an F-center, so named from the German word Farbenzentrum which translates to color center. Figure 9.7 shows the schematic of such a defect. Note that in the crystal, normally an ionic material, an anion is vacant from a position, and an electron occupies the space. A phenomenon associated with this type of defect is that the electron in the vacancy absorbs light. Thus, a crystal that was colorless becomes colored, with the color differing with different atoms, and the color intensity increasing with increasing numbers of F-centers. The alkali halides have been studied extensively in terms of these defects, since they are normally without color, and since they take on color under various forms of irradiation.

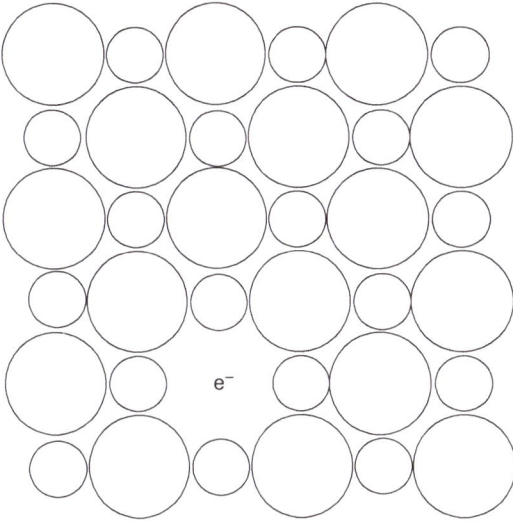

Figure 9.7: F-center defect.

9.4 Extended defects

As the name implies, extended defects are not localized to a point or points; rather they are created through some stresses to the material, or some condition during its growth. Such defects can be in two dimensions, along crystallographic planes, or can be three-dimensional, cutting through more than two planes.

What are termed Wadsley defects are those caused by a missing plane of one type of atom. Figure 9.8 shows this in three parts. To the left, Figure 9.8a shows the normal plane of two types of atoms or ions. The center of Figure 9.8b shows the absence of one type of atom along a plane perpendicular to that shown (sticking out of the paper and going back below it). To the right, Figure 9.8c shows the rearrangement of the atoms to restore order, but above and below the plane the octahedra must share edges. This is the shear plane.

As Figure 9.8 illustrates, Wadsley defects are a zone of relatively high numbers of defects that occur in specific directions in a crystal. These are sometimes called shear plane defects, and can be caused by different stresses on a crystalline material, such as repeated heating and cooling.

Overall, the presence of defects of any type within an otherwise crystalline material is related to the increase in entropy (ΔS) that is present as a material becomes less ordered. Increasing entropy equates to an increase in disorder; and thus the formation of defects is energetically favorable.

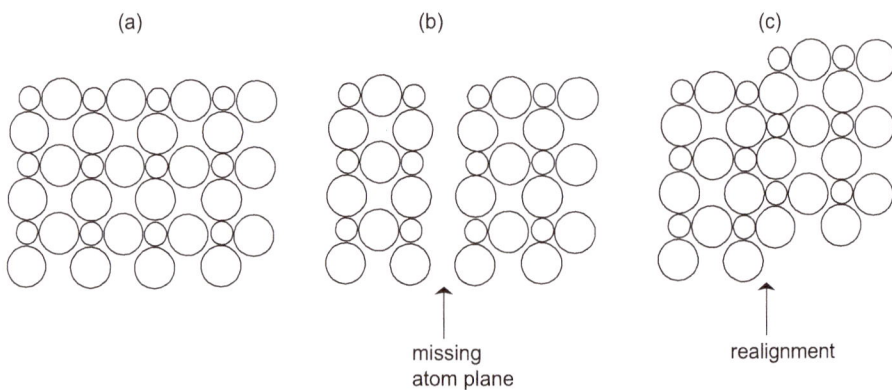

Figure 9.8: Wadsley defects.

References

[1] H. Gengxiang, X. Cai, Y. Rong. Materials Science: Volume 1, Structure, 2021, Walter DeGruyter GmbH.
[2] W. Neumann, A. Mogilatenko, K. Scheerschmidt. The Nature of Crystal Defects: Formation, Structure, and Analysis, 2022, Walter DeGruyter, GmbH.
[3] W. Neumann, A. Mogilatenko, K. Scheerschmidt. 2022, Walter DeGruyter, GmbH. ISBN: 978-311062145-7.
[4] W.H. Schottky, *Thermodynamik*, 1929.
[5] W.H. Schottky, *Physik der Gluhelektroden*, 1928.

Chapter 10
Conductors

10.1 Introduction of metal-based conductors

One common phenomenon of all metal elements is that they conduct electricity. To casual observation, there appears to be no difference between the rates at which one metal and another conduct. For example, both copper and aluminum can be used for wiring, since both conduct electricity at a rate that has proven useful for many types of wiring. There are measurable differences, however, and because of these, different metals are used in different, specific applications. The reason or reasons for this may be based on performance, may be based on economics (the cost of the material and how easy it is to form into desired shapes), or may be based on some combination of both. For example, gold contacts have found use in computers because only small amounts are required, because gold's higher conductivity than most other metals makes it useful in such an application, and because gold is highly resistant to corrosion. Copper is used for electrical wiring in homes, public buildings, and a myriad of other structures because of the low cost and easy availability of the metal when compared to other metals (such as gold). Table 10.1 lists several metals and their conductivity.

Table 10.1: Metals and conductivities.

Element	Conductivity (S/m)	Comments
Aluminum	3.5×10^7	Widely used based on cost
Copper	5.98×10^7	
Copper, annealed	5.80×10^7	Widely used based on cost
Gold	4.52×10^7	Used in environments where corrosion must be minimized
Iron	1.04×10^7	
Tungsten	1.82×10^7	
Silver	6.30×10^7	
Zinc	1.68×10^7	Alloyed with lead or tin to produce solders

Silver has been found to be the most conductive of all the metals, as seen in Table 10.1. The reason it has not been used extensively in wiring and other industrial applications is simply a matter of cost coupled with the effectiveness of copper. Put bluntly, copper is the cheaper metal of the two, and does the job of moving electrons to provide electrical current effectively enough.

When the overall mass is a factor in the development of some electrically conducting system, aluminum wiring has routinely been a material of choice. This is because its density, 2.7 g/cc, is far lower than that of silver, 10.49 g/cc, or of copper,

https://doi.org/10.1515/9783112205822-010

8.96 g/cc. In the 1950s, aluminum wiring was used extensively in home wiring. This has been largely discontinued because of the hazard of fire associated not with the actual aluminum wiring, but with the connections at the end of wires. Correcting this in residences usually involves the addition of a short piece of copper wiring between the aluminum wire and a switch or terminal, which is far more economically advantageous than removing and reinstalling entire wiring systems.

10.2 Nonmetal elements and conductivity

We will discuss semiconducting materials in Chapter 11. Several elements and mixtures conduct electricity less well than metals, but well enough that the controlled movement of electrons through the material can be useful.

We will discuss the phenomenon of superconductivity in Chapter 12.

Graphite (C_6) is the allotropic form of carbon that remains the only nonmetallic element which has been found to conduct, and that can be classified as a conductor of electricity. With an electrical conductivity of approximately 2×10^5 S/m, the free electron in each carbon atom of graphite enables this level of conductivity.

Several of the elements along the metalloid line conduct electricity to some degree and can be classified as semiconductors under certain conditions and conductors at others.

10.3 Organic conductors

Several polymers are conductive to some extent. Called intrinsically conducting polymers (ICP), Table 10.2 lists several of the most common. Two, polyacetylenes (PAC) and poly(p-phenylene vinylene) (PPV), have over time become the most extensively used. Polythiophene has also found several niche commercial applications, including as biosensors and as light-emitting diodes (LEDs).

Perhaps the obvious common factor in the conductive polymers shown in Table 10.2 is their extreme, high levels of conjugated unsaturation. As a point of comparison, polyethylene has no unsaturated sites in it, and is not conductive. But even polyacetylene, which contains no phenyl rings, has enough conjugated unsaturation that it is conductive. Molecules such as polyfluorene are heavily unsaturated in each repeat unit. And molecules like polynaphthalene can be formed in different isomers, with different carbon atoms in the rings acting as the bridging atoms from one naphthyl unit to another.

Table 10.2: Conducting polymers.

Name	Abbreviation	Repeat unit, Lewis structure	Comments
Polyacetylene	PAC		Doping increases conductivity dramatically. Can be cis or trans
Poly(*p*-phenylene vinylene)	PPV		
Polyazulene			Potential for use as thermoelectrics
Polyfluorene			Can be photoluminescent
Polynaphthalene			Different isomers exist
Polyphenylene sulfide	PPS		Semiconducting when doped
Polypyrene			Higher conductivity at elevated temperatures. Exhibits fluorescence
Polythiophene	PT		Varieties are used in LEDs

10.3.1 Doped polyacetylene

Polyacetylene in film form can be doped with p-dopants, in this case meaning substances such as Cl_2, which are capable of extracting an electron from the atoms of the main carbon chain. Other materials that have been found to do this include the other heavy halogens, as well as arsenic pentafluoride. The mechanism whereby conductivity occurs is generally thought to be because charge-transfer complexes are established between the halogen and the main chain of the polymer. Essentially the polymer acts as a long chain cation with the introduced halogen acting as the anion.

Logically, if p-dopants can be used to produce conducting polyacetylene, n-dopants should function in this manner as well, but in reverse as far as ionic charge. This means the polymer should now act as an anion, while the introduced material – such as the alkali metals lithium, sodium, and potassium – functions as the cation. It has been observed that n-dopants are sensitive to both moisture and air, which mimics the way these elements act when they are reduced metals.

There are differences between the conductivity of the two different isomers of polyacetylene, meaning *cis-* and *trans-*polyacetylene. Doped *cis-*polyacetylene generally can achieve conductivities roughly twice that of the trans-isomer. Doping with either bromine or iodine, for example, increases the conductivity of the material by several orders of magnitude.

10.3.2 Poly(*p*-phenylene vinylene)

Much like polyacetylene, the poly(*p*-phenylene vinylene) – PPV – is not conductive when it is not doped [1]. It is actually diamagnetic when undoped, but can be processed into thin films which can then have a dopant introduced. When doped with sulfuric acid (H_2SO_4), its conductivity can rise to as high as 100 S/m. Dopants vary considerably, with both the just mentioned sulfuric acid as well as triflic acid (CF_3SO_3H) having been used successfully. But the following have also found use as dopant materials in poly(*p*-phenylene vinylene): lithium, sodium, potassium, ferric chloride, and iodine.

10.3.3 Conductive metal-organic frameworks (MOFs)

Metal-organic frameworks, or MOFs, were mentioned in Chapter 5 as a hybrid type of material that requires both an inorganic and an organic component to create an end material which has a large, stable, open, central cavity or pore. They have evolved quickly into an entire class of materials, but have in common cube-like structures with metal ions in the 8 corners, connected with coordination complex bonds (dative bonds) to organic moieties, and rigid organic spacers that make up the 12 edges.

More recently, MOFs [2–7] have been found to be conductive, and, according to Xie and coauthors [3], conduct through four different means of transport. They are:

1. Through bond approach – a series of linked coordination compound bonds which are able to move electrons from site to site
2. Extended conjugation approach – delocalized systems on a large scale are formed between the metal centers and ligands
3. Through space approach – alignments of organic components permit pi-pi stacking interactions through phenyl rings, and electron movement via such stacking
4. Guest promoted approach – as the name implies, some "guest" component interacts in the large, empty space inherent in the MOF [3]

The diagram of a MOF shown in Figure 10.1a is an idealized schematic. The round corners, as expanded in Figure 10.1b can be any number of metal ions at the center of a moiety which can act as a corner, and connect to an organic spacer. The line, as expanded in Figure 10.1c has originally been xylyl units in some of the earliest MOFs, but now can be any number of organic pieces, provided they give enough rigidity to the structure that the large, central cavity remains stable.

The potential applications for conductive MOFs are still emerging, and at the present none have become widely or commercially available. The amount of research that has been done on this novel class of materials, however, certainly indicates that applications are on the near horizon.

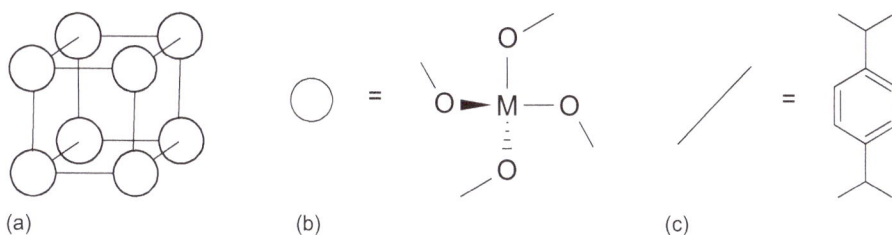

(a)   (b)   (c)

Figure 10.1: MOF schematic.

References

[1] J. Banerjee, K. Dutta. A short overview on the synthesis,properties and major applications of poly(p-phenylene vinylene, *Chemical Papers*, 2021, 75: 5139–5151.
[2] M. Jacoby. MOFs have a new trick, *Chemical and Engineering News*, November 29th 2021, 30–34.
[3] L.S. Xie, G. Skoupskii, M. Dinca. Electrically conductive metal-organic frameworks, *Chemical Reviews*, 2020, 120(16): 8536–8580.
[4] Z. Meng, A. Aykanat, K.A. Mirica. Welding metallophthalocyanines into bimetallic molecular meshes for ultrasensitive, low-power chemiresistive detection of gases, *Journal of the American Chemical Society*, 2019, 141(5): 2046–2053.

[5]   M. Ko, L. Mendecki, A.M. Eagleton, C.G. Durbin, R.M. Stolz, Z. Meng, K.A. Mirica. Employing conductive metal-organic frameworks for voltammetric detection of neurochemicals, *Journal of the American Chemical Society*, 2020, 142(27): 11717–11733.

[6]   C. Park, W.-T. Koo, S. Chong, H. Shin, Y.H. Kim, H.-J. Cho, J.-S. Jang, D.-H. Kim, J. Lee, S. Park, J. Ko, J. Kim, I.-D. Kim. Confinement of ultrasmall bimetallic nanoparticles in conductive metal-organic frameworks via site-specific nucleation, *Advanced Materials*, 2021. https://doi.org/10.1012/adma. 202101216

[7]   H. Zhong, M. Ghorbani-Asl, K.H. Ly, J. Zhang, J. Ge, M. Wang, Z. Liao, D. Makarov, E. Zschech, E. Brunner, I.M. Weidinger, J. Zhang, A.V. Krasheninnikov, S. Kaskel, R. Dong, X. Feng. Synergistic electroreduction of carbon dioxide to carbon monoxide on bimetallic layered conjugated metal-organic frameworks, *Nature Communications*, 2020. doi: 10.1038/s41467-020-15141-y

Chapter 11
Semiconductors

11.1 Introduction

While elemental metals and alloys are usually very good conductors, nonmetallic elements, organic molecules, and highly ionic compounds tend to be nonconductive. Certain elements and mixtures that have been developed and exploited precisely because they conduct, but do not do so as well as metals, are called semiconductors. Semiconductors were developed in the past century precisely because they are an essential part or component of many electrical user-end items. The semiconductor industry is large enough worldwide that there are now multiple trade organizations devoted to their uses and for advocating such materials and their potential [1–4]. Those we discuss in this chapter have become central to many electronic devices.

In discussing semiconductors, the term bandgap inevitably becomes important in any explanation. Sometimes also called the energy gap, this is a range of energies within a solid material in which there are no electronic states. A diatomic molecule has discrete electronic levels of bonding and antibonding electrons; but when millions of atoms are examined in a material, such discrete states can be said to blend into a band of energy. The lower limit of the bandgap for a material is the upper value of the valence band (the band occupied by electrons) and the upper limit is the lowermost value of the conduction band (the band to which electrons can be promoted so that current does flow). The valence band is that zone of energy in which discrete molecules contain electrons that do not move through a material. The conduction band is, as mentioned, a zone of energy in which electrons can move through a material making that material in some way conduct electricity. And the controlled movement of electrons through different materials is the heart of what any semiconducting material does. Figure 11.1 shows the graphic distribution of electrons in these bands for conducting, semiconducting, and nonconducting materials.

Notice that in any insulator, shown to the right of Figure 11.1, the difference in energy between the valence band and the conduction band – the bandgap – is large enough that electrons cannot be promoted from one to the other by heat or light. Note also that for materials which are conductors, shown to the left of Figure 11.1, the valence and conduction bands virtually overlap, which means electrons can easily flow through the material, since there is no gap to overcome. The semiconductor materials are thus those for which there exist an energy difference between the valence band and the conduction band, but for which the gap is small enough in energy that electrons can be promoted from one to the other. A full listing of the bandgaps for all semiconductors would be quite large. Table 11.1 shows some of the most common semiconductor materials and their bandgaps.

https://doi.org/10.1515/9783112205822-011

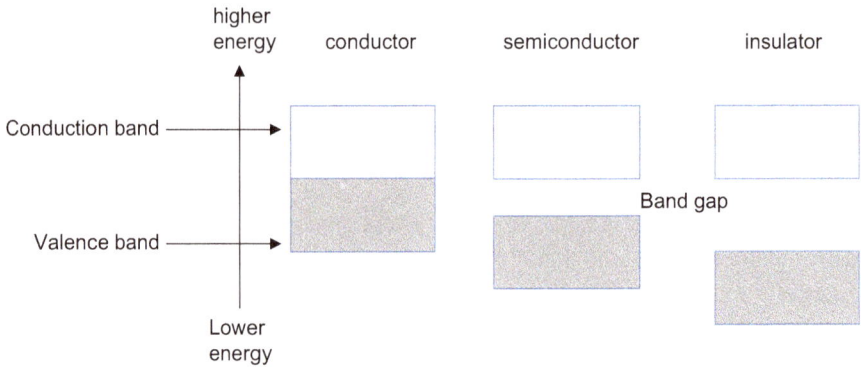

Figure 11.1: Bandgap in three classes of materials.

Table 11.1: Common semiconductor material bandgaps.

Name	Formula	Bandgap (eV)	Comments
– Element			
Carbon (C_{10}) diamond	C	5.5	Often lab-grown diamonds
Germanium	Ge	0.67	Rarer material than silicon
Selenium	Se	1.8	Photoconductor
Silicon	Si	1.14	Very common material for semiconductors
Tellurium	Te	0.3	Photoconductor
– Compound			
Aluminum nitride	AlN	6.0	Extremely large bandgap, intrinsic semiconductor
Aluminum phosphide	AlP	2.5	–
Cadmium telluride	CdTe	1.6	Relatively new semiconductor
Copper(I) oxide	Cu_2O	2.1	–
Gallium antimonide	GaSb	0.73	–
Gallium arsenide	GaAs	1.43	–
Gallium nitride	GaN	3.4	Has very high thermal conductivity
Gallium oxide	Ga_2O_3	4.66	–
Gallium phosphide	GaP	2.26	–

Table 11.1 (continued)

Name	Formula	Bandgap (eV)	Comments
Indium phosphide	InP	1.34	–
Lead(II) sulfide	PbS	0.37	One of the earliest semiconductors
Silicon nitride	Si_3N_4	5.0	–

Interestingly, there remains some debate about the value of the largest bandgap that can be considered a workable one for a semiconductor. Some sources indicate that gallium oxide (Ga_2O_3) is the largest bandgap for any workable semiconductor, but we include some materials in Table 11.1 that have higher bandgaps because they can find uses in what can be considered extreme conditions, such as extremely high pressure.

The means by which conduction takes place in a semiconductor is that heat or light promotes an electron to move through the material by adding energy that moves it from the conduction band over the bandgap to the conduction band. When there is a slight imbalance of extra electrons in the material, it is stated that these electrons are promoted to the conduction band and can then move through the material. When the imbalance is toward slightly less electrons in the material, it is often said that positive holes move through the material. To be clear, it is still electrons that move, but the hole – the positively charged space caused by an electron's movement away from that space – is filled by another electron, thus apparently making a new hole. When this is done countless times, it appears that the positively charged holes are what are moving.

11.2 Silicon

The heart of many semiconductors is the element silicon, after it has been purified to a high degree. The means by which silicon is refined from silica (SiO_2) to high purity is through a zone refining furnace, the schematic of which was shown in Figure 9.1. This method of purification is expensive, and thus another means of forming silicon into wafers for use in semiconductor devices is sometimes used, in which silane (SiH_4) is used as a starting material, and at elevated temperatures silicon is deposited on a surface. This tends to leave small traces of Si–H behind, which in the right environment can enhance the semiconducting properties of the material.

Silicon is one of several intrinsic semiconductors. This means that it is conductive without any additive, known as a dopant. Elements can be doped into a semiconductor in a controlled manner – added in some small amount to the element silicon (or other semiconducting elements) – to influence the flow of electrons which was just described. The dopant often adds an electronic layer that lowers the bandgap in the otherwise undoped material. The addition of a dopant makes the resulting material an extrinsic semiconductor.

To be more precise, the term doping indicates the addition of a small amount of some material to an already semiconducting material, but specifically an addition to adjust how well the resultant material conducts. Routinely this involves a specific increase in conductivity. This in turn means that the addition of an element must be one that either increases or decreases the amount of electrons in the target material.

The term p-doping refers to the inclusion of a very small amount of an element, such as boron or gallium, which is deficient in a valence electron when compared to the semiconductor material. What is called an n-dopant is an element such as phosphorus or arsenic that is present in a semiconductor, and that adds a slight excess of electrons to the material. It is important to realize that even small amounts of a dopant can increase the conductivity of the resulting semiconductor by orders of magnitude. Figure 11.2 shows stylized examples of each type of dopant.

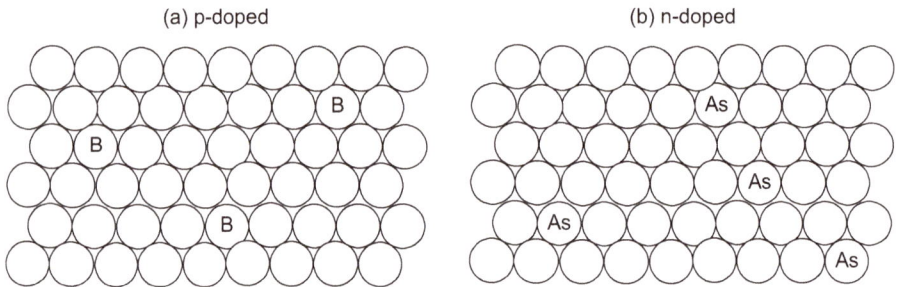

Figure 11.2: Silicon with p- and n-dopants.

11.3 Germanium

Colocated on the periodic table of the elements with silicon, directly below it, germanium also has found extensive use as a semiconductor. The bandgaps for both germanium and silicon are given in Table 11.1. Curiously, the earliest semiconductors were virtually all germanium, when the industry was not nearly as large as it is today.

Germanium is one of only a handful of elements that were unknown at the time when Dmitri Mendeleev postulated the first periodic table, but which he predicted should exist, inclusive of several of its properties. Germanium is never mined and refined as a single element; rather, it is a by-product of zinc production, from minerals such as sphalerite, normally a zinc, iron sulfide (usually written (Zn,Fe)S, since germanium is only a trace). The chemistry by which elemental germanium is isolated is complex, but can be represented as follows, in Figure 11.3.

The third reaction seems to be a return to the starting point, but the germanium oxide produced in this step is much purer than the original starting material, and is

$GeS_2 + 3O_{2(g)} \rightarrow \Delta \rightarrow GeO_2 + 2SO_{2(g)}$ (along with zinc oxides)

then

$2Cl_{2(g)} + GeO_2 \rightarrow GeCl_4 + O_{2(g)}$

then

$GeCl_4 + 2H_2O \rightarrow GeO_2 + 4HCl$

then

$GeO_2 + 2H_{2(g)} \rightarrow Ge + 2H_2O$

Figure 11.3: Germanium refining.

free of zinc-included impurities. From this point, hydrogen can be used as a reducing agent, with elemental germanium being the final product.

The total amounts of germanium produced each year are rather small, with the US Geological Survey Mineral Commodity Summaries (the MCS) recording quantities in kilograms, whereas many metals and other commodities are recorded in tons or even thousands of tons. The MCS states, "A company in Utah produced germanium wafers for the semiconductor industry and for solar cells used in satellites from imported and recycled germanium," but indicated that germanium was produced from zinc concentrates in both Alaska and Tennessee [5]. The fact that germanium is refined to the purity level needed for solar cells only from the concentrates of other elements indicates how scarce this element is.

The means by which electrons move in germanium is the same as that we have just explained for silicon semiconductors. The only real difference is the bandgap, both of which are listed previously, in Table 11.1.

11.4 Gallium arsenide

Another material used extensively as a semiconductor is gallium arsenide (GaAs). It is so prevalent within the semiconductor industry that the gallium required to produce this type of semiconductor is tracked by both the United States Department of Defense and the US Department of Energy [6, 7]. Elemental arsenic is known to occur, but the element is also extracted from a wide variety of minerals. Realgar and orpiment are two minerals that yield it, but arsenic can also be reclaimed from the dusts associated with copper metal refining.

Like silicon and germanium, gallium arsenide is an intrinsic semiconductor. In this case, it is because the same number of atoms exist within it that are deficient by one electron (Ga) as there are atoms with one extra electron (As). More than one method exists whereby the compound is made. They include:
1. The Bridgman–Stockbarger technique: a technique that has been used widely for growing different types of crystalline materials. For gallium arsenide, the ele-

ments react at elevated temperature in a furnace, and grow as a deposited solid on a seed crystal in a relatively cool portion of the furnace.

2. Vertical gradient freeze method (VGF): the predominant means by which GaAs wafers are manufactured. In this technique, starting material and a seed crystal, in a crucible, are heated to melt, then cooled in a controlled manner so that the crystal grows in the direction of the cooling. Crystals are then sliced to the proper sizes and smoothed. The technique produces wafers with extremely low numbers of defect sites.

3. Liquid encapsulated Czochralski growth (LEC): a technique that can be used to grow several different types of semiconductors, including those of elemental silicon. By dipping a seed crystal into the molten element or compound and extracting it slowly, large cylindrical crystals can be grown.

4. Vapor reaction of gallium and arsenic trichloride. The simplified reaction chemistry is: $2Ga + 2AsCl_3 \rightarrow 2GaAs + 3Cl_2$. Elemental chlorine must be recovered, but can be reused.

5. Metalorganic chemical vapor deposition (MOCVD) of trimethyl gallium and arsine, according to the reaction: $AsH_3 + Ga(CH_3)_3 \rightarrow 3CH_{4(g)} + GaAs$

11.5 Other non-silicon-based semiconductors

11.5.1 Metal oxide semiconductors

Metal oxide semiconductors (MOS) are another class of materials that has been used widely in the semiconductor industry. Discovered and first examined in the 1960s, MOSs can also be divided into n-type and p-type materials, as well as inherent semiconductors. Once again, a full listing would be quite large, but some of those that have been studied in detail are shown in Table 11.2.

Table 11.2: Metal oxide semiconductors.

Name	Formula	Bandgap (eV)	Comments
– n-type			
Tin(IV) oxide, or stannic oxide	SnO_2	3.6	–
Zinc oxide	ZnO	3.37	High mechanical and chemical stability
Titanium dioxide	TiO_2	3.2	–
Tungsten trioxide	WO_3	2.6–3.0	Also used as a pigment and in paints

Name	Formula	Bandgap (eV)	Comments
– p-type			
Cobalt(II,III) oxide	Co_3O_4	1.6–2.2	–
Nickel(II) oxide	NiO	3.6–4.0	Wide bandgap semiconductor

One of the prevalent uses for metal oxide semiconductors is the integrated circuit. Shown in two states in Figure 11.4, these are often abbreviated as MOFSET for metal oxide semiconductor field effect transistor. A basic explanation of how they function can be broken down into the following steps:

1. Shown is one – an integrated circuit – of a large, ordered array. Often these are transistors.
2. Transistors enable precise current flow with two possible functions: (1) amplify a signal and (2) turn current on or off (becoming an on/off switch)
3. One type is a semiconductor doped with two regions of the other – for example, a p-type silicon semiconductor with two n-type regions of silicon.
4. A source and a drain are needed, in contact with the n-type.
5. A metal oxide, such as SiO_2, is then layer deposited on top.
6. A third electrode is atop the SiO_2 layer – called the gate because of its function.
7. A potential can be applied to the gate.
8. The applied potential allows electrons to migrate under the SiO_2 layer between the two electrodes – as shown in Figure 11.4b; a channel is formed.
9. The gate can have two possible functions: (1) control amount of current or (2) turn potential on and off, becoming an on/off switch

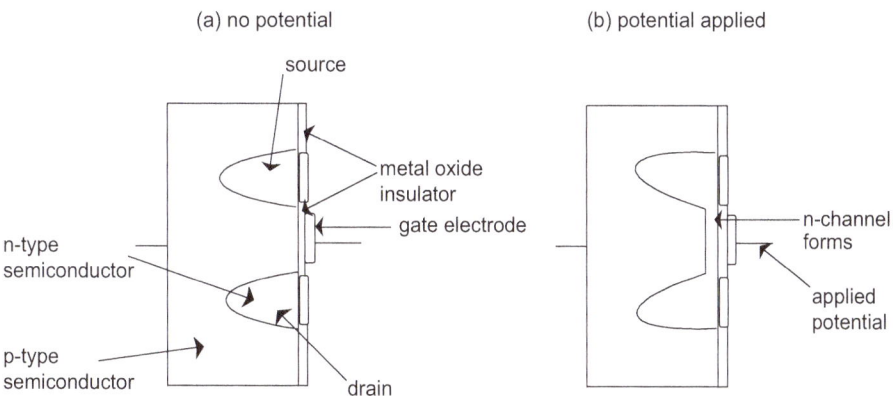

Figure 11.4: Metal oxide semiconductor field effect transistor (MOFSET).

11.5.2 Semiconducting polymers

We have mentioned in Chapter 10 that certain polymers can be considered conductors. When the dopant level is correct, several polymers act as semiconductors, some of them being the same in polymeric structure as the conductors. We discuss here three that have been investigated for decades.

1. Polyacetylene

Several repeat units of polyacetylene are shown in Figure 11.5, in both its *cis-* and *trans-* isomer configurations. The material can be difficult to form, and early attempts to do so directly from acetylene were not successful. Modern methods often involve some ring-opening reaction. To bring the material to a semiconducting state, it must be doped with arsenic pentafluoride (AsF_5), molecular iodine (I_2), or other molecules, and unfortunately still suffers from being easily oxidizable with molecular oxygen.

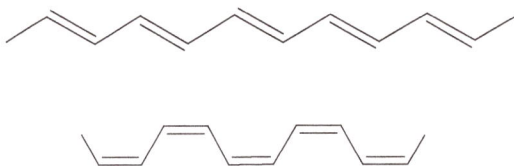

Figure 11.5: Polyacetylene.

2. Poly(*p*-phenylene)

Poly(*p*-phenylene) is a rigid polymeric material because of its repeating phenyl structure, illustrated in Figure 11.6. Traditionally, it has proven difficult to fabricate, because the chain length does not routinely form in a uniform manner. Interestingly, although the structure for poly(*p*-phenylene) can be drawn as a string of phenyl units laying on the same plane, the steric hindrances of hydrogen atoms *ortho* to the phenyl-to-phenyl carbon–carbon bond tilts the rings out of planarity. If this is considered a deformation, it is one that raises the bandgap, but still permits doped versions of the material to behave as semiconductors. Also, it has proven to be chemically more robust than polyacetylene.

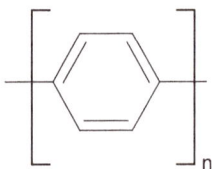

Figure 11.6: Poly(*p*-phenylene).

3. Polyphenylene vinylene (PPV)

Somewhat like polyacetylene, PPV, as shown in Figure 11.7, has limited solubility in various solvents, which makes formation of films challenging. This can be alleviated by altering side chains to the main polymer chain, but this tends to change the bandgap. Altering the side chains in such materials changes the ability of the repeat units to "lay flat" or be planar, which in turns lowers the range over which the pi-system can function as a single, extended system.

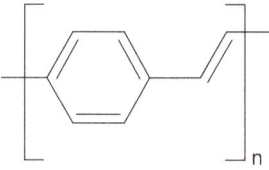

Figure 11.7: Polyphenylene vinylene.

In these three organic materials, and in many others, there is concern and effort to design materials that have planar configurations, meaning the entire extended pi-system is flat and presumably able to delocalize electrons throughout. The traditional means of producing such polymers involves covalent bonding between different rings. In recent years, the perceived limitations of such synthetic strategies has been addressed by employing what are called "noncovalent through-space intramolecular interactions" [8], which have been successful in the case of some polymers, resulting in higher regions of planar, extended pi-systems [8].

These and other polymers that can semiconduct routinely require a dopant, and the choice can be one of many. Species as different as elemental iodine (I_2), bromine (Br_2), and ferric chloride ($FeCl_3$) have been used successfully. These dopants tend to partially oxidize the polymer, promoting electrons into the conduction band.

The field of semiconducting polymers continues to grow and evolve, with applications continuing to grow as well. In an interesting note, the mechanism by which polymers can act as semiconductors continues to evolve as well, as there is not yet universal agreement upon the mechanism, or mechanisms, that explains the promotion of electrons to the conduction band, and especially that explains how electrons move from one polymer chain to another.

11.6 Recycling

Despite the wide use of semiconductors in a very large array of devices – production which includes hundreds of millions of devices per year – the size of each device is small enough that no national or regional recycling programs exist for them. Organizations that track the recycling of many materials usually track what is called e-waste as well, meaning discarded materials that come from electronics. Of the millions of tons of e-waste that are generated annually, less than 20% are collected in some manner that is different from direct disposal, and recycled [9–11].

References

[1]   Semiconductor Industry Association. Website. (Accessed 16 June 2025, as: https://www.semiconductors.org).

[2]   Global Semiconductor Alliance. Website. (Accessed 16 June 2025, as: https:www.gsaglobal.org).

[3]   Canada's Semiconductor Council. Website. (Accessed 16 June 2025, as: https://www.canadassemiconductorcouncil.com).

[4]   Euseimconductors.eu. Website. (Accessed 16 June 2025, as: https://www.eusemiconductors.eu).

[5]   United States Geological Survey. Mineral Commodity Summaries 2024, Downloadable as: pubs.usgs.gov/publication.mcs2024).

[6]   Strategic and Critical Materials 2013 Report on Stockpile Requirements. Downloadable as: https://mineralsmakelife.org.Strategic_and_Critical_Materials_2013_Report_on_Stockpile_Requirements.pdf

[7]   U.S. Department of Energy. Critical Materials Strategy, downloadable at: https://www.energy.gov/eere/ammto/2021-doe-critical-materials-strategy).

[8]   H. Huang, L. Yang, A. Facchetti, T.J. Marks. Organic and polymeric semiconductors enhanced by noncovalent conformational locks, *Chemical Reviews*, 2017, 117: 10291–10318. and references therein.

[9]   Azo Materials, How Do We Recycle Semiconductors. Website. (Accessed 10 June 2025, as: https://www.azom.com).

[10]   Global e-Waste Monitor 2024. Website. (Accessed 16 June 2025 as: https://www.itu.int).

[11]   The Global e-Waste Statistics Partnership. Website. (Accessed 16 June 2025, as: https://globalewaste.org).

Chapter 12
Superconductors

12.1 Introduction

The idea of a superconductor is rather simple in theory: a material that superconducts has zero resistivity to the flow of electric current at or below a certain temperature. It also must expel a magnetic flux field, and again does so at and below a specific temperature, one unique to each material. This phenomenon is therefore a function of both temperature and magnetic field. This phenomenon occurs in certain materials – in some elemental solids, as well as a wide variety of solid compounds. From its initial discovery in 1911, this routinely has been found to occur at a very low absolute temperature for most of the materials that manifest the phenomenon.

All superconductors have what is known as a critical temperature, or a transition temperature below which the phenomenon occurs, noted T_c. Below this temperature, resistance becomes zero, with a dramatic drop being a hallmark of those superconducting materials that were identified in early research into the field. Above that temperature, the conductivity of each material is different. Normal conductors do not have this dramatic, critical point, but rather have a slow, gradual loss of resistivity as the temperature of the material is lowered.

Also, even when a material is cooled lower than the T_c, if a large enough magnetic field is applied, superconductive behavior will cease. This is the critical field, H_c. The critical field is thus dependent upon temperature as well, and decreases as the temperature of any material increases.

12.2 Known elemental superconductors

The first recorded case of superconductivity in any material was in solid, elemental mercury. In 1911, Heike Kammerlingh Onnes noticed that when cooled to 4 K, in liquid helium, all resistance to electrical conductivity in solid mercury was gone. This temperature was difficult to produce in 1911, and remains one that can only be reached with specialized equipment today, as well as with liquefied helium as a coolant. Since that time, many elements, and some elements doped with a second (alloys or mixtures) have been found to superconduct. The transition temperatures or critical temperatures (their T_c) of these elements tend to be low, although successive reports are routinely of higher temperature superconductors, since this idea of high temperature superconductivity is one of the desired goals when studying novel materials for their superconducting potential. Table 12.1 shows a list of elements that exhibit superconductivity, with their T_c values listed in ascending order.

https://doi.org/10.1515/9783112205822-012

Table 12.1: Superconducting elements.

T_c (K)	Symbol	Name
0.000325	Rh	Rhodium
0.0004	Li	Lithium
0.00053	Bi	Bismuth
0.015	α-W	Tungsten
0.14	Ir	Iridium
0.165	Hf	Hafnium
0.39	Ti	Titanium
0.49	Ru	Ruthenium
0.52	Cd	Cadmium
0.55	Zr	Zirconium
0.65	Os	Osmium
0.68	α-U	Uranium
0.855	Zn	Zinc
0.92	Mo	Molybdenum
1.0	β-W	Tungsten
1.083	Ga	Gallium
1.20	Al	Aluminum
1.37	α-Th	Thorium
1.4	Pa	Protactinium
1.8	β-U	Uranium
2.39	Tl	Thallium
2.4	Re	Rhenium
3.4	In	Indium
3.72	Sn	Tin
3.95	β-Hg	Mercury
4.15	α-Hg	Mercury
4.48	Ta	Tantalum
4.9	α-La	Lanthanum
5.03	V	Vanadium
6.3	β-La	Lanthanum
7.19	Pb	Lead
7.46	Tc	Technetium
9.26	Nb	Niobium

As can be seen, while a significant number of elements do show superconductivity at extremely low temperatures, none do so at a temperature that is high enough that applications for it will be economically feasible in the near future. As well, some of the elements in Table 12.1 are rare enough (such as technetium or niobium) that it would be virtually impossible to scale up their production and use in a significant way for any application that might be found for them.

To provide some perspective of what elements do and do not superconduct, not as a function of critical temperature, Figure 12.1 shows a periodic table of the elements, displaying symbols only for those elements which have be shown to display superconductivity at any temperature.

Li	Be											Al	Si	P			
												Al	Si	P			
		Ti	V	Cr						Zn	Ga	Ge	As	Se			
	Y	Zr	Nb	Mo	Tc	Ru	Rh		Cd	In	Sn	Sb	Te				
Cs	Ba	Hf	Ta	W	Re	Os	Ir		Hg	Tl	Pb	Bi					

	La	Ce				Eu						
	Th	Pa	U			Am						

Figure 12.1: Periodic table noting superconducting elements.

Even seeing the periodic table in this manner does not immediately clear up why some elements superconduct and others do not. It is tempting, for example, to claim that elements at the beginning of shells, such as p, d, and f shells, must have electron configurations and enough unpaired valence electrons that they are favorable for superconductivity. Yet this ignores the fact that zinc, cadmium, and mercury, as well as beryllium and barium, display the phenomenon.

The elements that are superconducting are often called type I superconductors, those that exhibit this at the lowest temperatures. What are called type II superconductors are often alloys or compounds – in many cases, ceramic compounds – which can superconduct, and which do so at considerably higher temperatures.

The mechanism by which elements, and compounds – which we will discuss next – superconduct remains a subject of discussion and debate. What is called the BCS theory, for Bardeen–Cooper–Schrieffer, is the oldest theory attempting to explain the phenomenon – and for which John Bardeen, Leon Cooper, and John R. Schrieffer received the 1972 Nobel Prize for physics. The basic explanation of the theory, broken into steps, includes:

1. Electrons moving through the material cause slight deformations in the nuclei of the lattice at the atoms closest to which the electron passes. This creates an increased positive charge density.
2. The deformation caused by an electron in turn causes a second electron to be attracted to the area that now has a positive charge density.
3. Electron–electron repulsions that normally exist in a material are then overcome at these extremely low temperatures and areas of enhanced positive charge.
4. The two electrons come together in what are called Cooper pairs.

5. At low temperatures, the collective Cooper pairs remain together, resist localization back into the lattice, and the resultant flow of electrons takes place with no resistance.
6. This indefinite flow of electrons continues, and in the process generates a strong magnetic field.

It has been pointed out however that this theory does not adequately explain why superconductivity occurs at higher temperatures, where a lattice returns to its normal state in a much shorter time.

12.3 Temperature correlations

One of the long-term goals of research into superconducting materials is to find and produce a superconducting material with a critical temperature at or slightly above ambient temperature, meaning 25 °C. Professional organizations devoted to superconductivity routinely discuss and examine this [1–7]. The elements just listed in Table 12.1 all show superconductivity only at extremely low temperatures, where applications are not practical. However, in the past three decades, several thousand materials have been found to be superconducting, and the upper limit for the T_c of several of these materials has now reached above the boiling point of liquid nitrogen, 77 K (−196 °C).

Currently, $Hg_{0.2}Tl_{0.8}Ca_2Cu_3O$, which has a T_c of 139 K (−134 °C) at a single atmosphere of pressure is the material with one of the highest T_c. But research is constantly continuing to find materials with even higher critical temperatures, and also examines materials at extremely high pressures, where superconductivity may be exhibited at still higher temperatures. Table 12.2 lists several of the existing higher temperature superconductors. It should be noted however, that some of them display the phenomenon at the just-mentioned extremely high pressures, which again means they do not represent materials that will feasibly find uses and applications in any widespread way.

Table 12.2: Examples of high-temperature superconductors.

Material	T_c in °C	T_c in K	Comments
LaH_{10}	−23	250	Pressure of 170GPa or 2.46×10^7psi
H_2S	−70	203	Pressure of 155GPa or 2.25×10^7psi
$Hg_{0.2}Tl_{0.8}Ca_2Cu_3O$	−134	139	At 1 atm
$Tl_2Ca_2Ba_2Ca_3O_{10}$	−148	125	At 1 atm [8]
$Bi_2Sr_2Ca_{n-1}Cu_nO_{2n+4+x}$	−163	110	Referred to as BSCCO, $n = 2$ most common
$YBa_2Cu_3O_{7-x}$	−180	93	Referred to as YBCO, or Y123
MgB_2	−234	39	Non-copper-containing

It is also worth mentioning that certain elements that are components of supercon-ducting compounds are relatively rare, and because of this are tracked by national organizations [8]. For example, thallium is tracked as being produced in kilograms per year, and the USGS Mineral Commodity Summaries states of it, concerning uses:

> Small quantities of thallium are consumed annually, but variations in pricing and value data make it difficult to estimate the value of consumption. The primary end uses included the follow-ing: radioisotope thallium-201 used for medical purposes in cardiovascular imaging; thallium as an activator (sodium iodide crystal doped with thallium) in gamma radiation detection equip-ment; thallium-barium-calcium-copper-oxide high-temperature superconductors [8]

Further, the theory of why superconductivity occurs at these elevated temperatures re-mains in debate. Since many of the structures of high-temperature superconductors ex-hibit some variation on the perovskite structure, as seen in simplified form in Figure 12.2, a significant amount of research has gone into making correlations between the phe-nomenon and the lattice structure. But since not all superconductors form in that struc-ture, it is felt that there must be more to any theory, so that it includes all higher temper-ature superconducting materials.

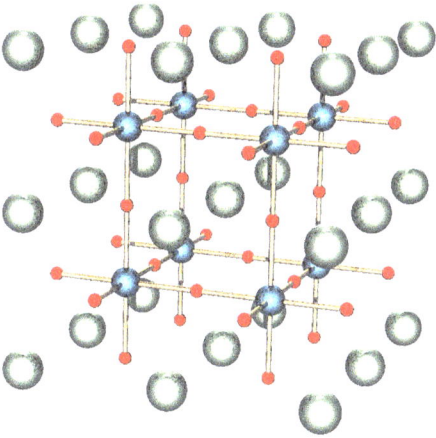

Figure 12.2: Perovskite structure.

Whatever the material is that superconducts though, for hopefully obvious reasons, any room-temperature (ca. 25 °C), superconducting end product could be put to a number of uses, creating new consumer items, improving the effectiveness of the flow of electricity through existing grids, bringing into existence a form of magnetic levitating train (mag-levs) that would save enormous amounts of energy, and overall, changing society quite dramatically.

Focusing on just one example, one that has been pursued both in terms of super-conductivity and through other means, a train that rides on superconducting magnets would represent astonishing savings in fuel for this form of transport. Once the train is put into motion and attains its normal traveling speed, there would be no need for further energy input. In theory, such a conveyance could travel to its destination using a tiny fraction of the energy that is required when compared to even the most energy-efficient trains in existence today, all of which must have a continuous input of energy to propel them forward.

The term "high-temperature superconductor" – often abbreviated HTS – may seem oddly used when it is pegged to 77 K (or −196 °C or −321 °F), but this number is chosen because it is the boiling temperature of liquid nitrogen. Liquid nitrogen can now be produced inexpensively, is a renewable resource, and is thus an inexpensive coolant. Liquid nitrogen is produced by the liquefaction and subsequent distillation of liquid air. The other products of this process are routinely liquid oxygen, liquid argon, and solid carbon dioxide, along with other trace materials such as krypton and xenon. Since it comes from air, liquid nitrogen is a renewable form of coolant, unlike liquid helium. Helium, which off-gases from our planet, is routinely extracted and concentrated from natural gas wells. It is localized enough that countries without it have gone to extremes to find means to obtain it. For example, the Chinese government has recently backed research into extracting helium from the surface of the Moon, where some accumulates after being ejected from the Sun.

12.4 Potential superconductors

Any material that is found to superconduct at any temperature is routinely tested at progressively higher temperatures until its T_c is found. The existing suite of high temperatures is such that there does not yet seem to be a hard, upper limit to where superconductivity ends. Numerous derivatives of known superconductors have been tried, and several reviews have been written about what is expected and hoped for among superconducting materials, as well as what has been achieved [9–11].

In 1987, the Nobel Prize was awarded to J. Georg Bednorz and K. Alexander Müeller "for their important breakthrough in the discovery of superconductivity in ceramic materials." It is not surprising that such an honor and recognition set off something of a rush in this already busy field, and made the expansion into a variety of mixed materials almost limitless. Numerous variations of the perovskite structure have been examined, although superconducting compounds such as magnesium boride appear to be proof that perovskites and copper oxides are not the only possibilities in this area.

Additionally, after researchers discovered a rich reaction chemistry in the fullerene class of carbon (the smallest fullerene being C_{60}), it was probably not a huge leap to examine them and their derivatives for any possible superconductivity. A group of M_3C_{60} materials, called metal fullerides, have been found to be superconducting. Admittedly, K_3C_{60} superconducts at or below 18 K, making it a low-temperature superconductor. But it is impressive to see how quickly such new materials were examined for this phenomenon (since the fullerene allotrope of carbon was only discovered and first reported in 1985).

12.5 Goals for superconducting materials

Perhaps obviously, one major goal for a future superconductor is that the material exhibits the phenomenon at room temperature, generally about 25 °C, as already mentioned.

Since that has yet to be achieved, two other relatively high temperatures that can be considered major goals for a superconducting material would be 0 °C (273 K), and −78 °C (195 K). These temperatures are the freezing points of water and carbon dioxide, two readily available materials that are very inexpensive coolants. Being able to uncouple the phenomenon from very low temperatures is widely considered to be the key that will open up a large number of applications for superconducting materials. Even though 77 K (−196 °C) is the temperature of liquid nitrogen, which is an inexpensive coolant, as mentioned, the −78 °C and 0 °C goals remain desired levels to reach because of the ease and simplicity of attaining those two temperatures.

References

[1]   American Superconductor (AMSC). Website. (Accessed 16 June 2025, as: https://www.amsc.com).
[2]   IEEE Council on Superconductivity. Website. (Accessed 16 June 2025, as: https://ieeecsc.org).
[3]   Superconductors.org. Website. (Accessed 16 June 2025, as: http://www.superconductors.org).
[4]   European Society for Applied Superconductivity. Website. (Accessed 16 June 2025, as: https://www.esas.org).
[5]   Institute of Technical Physics. Website. (Accessed 16 June 2025, as: https://www.itep.kit.edu).
[6]   IEEE Control Systems Society. Website. (Accessed 16 June 2025, as: http://australia-and-new-zealand.guide.ieeecsc.org).
[7]   Journal Novel Superconducting Materials, Walter DeGruyter GmbH.
[8]   U.S.G.S. Mineral Commoidty Summaries 2024, downloadable.
[9]   A.W. Sleight. Room temperature superconductors, *Accounts of Chemical Research*, 1995, 28: 103–108.
[10] S. Lemonick. Hunting for the next high-temperature superconductor, *ACS Central Science*, 2018, 4: 1447–1449.
[11] H. Kaur, H. Kaur, A. Sharma. A review of recent advancement in superconductors, *Materials Today: Proceedings*, 2021, 37: 3612–3614.

Chapter 13
Biomaterials

13.1 Introduction

Roughly, since the end of the Second World War, the entire world has been immersed in what history may eventually call an age of plastics. As discussed in Chapter 7, there exists a large array of plastics with a wide variety of properties, but only a few which are produced on such a large scale that they require codes of their own, known as Resin Identification Codes or RICs. The nature of these plastics or polymers is such that their bonds are robust enough that they are predicted to be stable for tens of thousands of years – a significant problem when their long-term disposal is considered. Additionally, since crude oil is the feedstock for the major, industrially significant plastics, there is a steadily rising concern that this resource is finite, and that the supply will be exhausted in possibly 50 years, and thus other feedstock materials should be found. This is the overall situation in which biomaterials become important, or bio-based materials do.

Somewhat like the term "materials chemistry," the term "biomaterials" has fuzzy edges. Some are of the opinion that this means exclusively materials that can be used in the human body for some medical purpose. Others widen the definition considerably and include materials that have a biological origin, but that are being produced to replace some material that is non-biological in origin, and that we perceive to have some vulnerability in terms of its long-term supply and sourcing. Biofuels are an excellent example of this, as are bioplastics. So are mycelium-based materials. Overall, there is enough interest and research efforts directed into areas that can be considered aspects of biomaterials that numerous trade organizations exist to promote them [1–4].

In this chapter, we will try to encompass the larger definition of biomaterials, at least in terms of those that are seeing wide and expanding uses. Additionally, we will note materials that are biological in origin, that were not originally designed to be medically applied, but that now find uses in the medical community.

13.2 Materials for medical use

Biomaterials are intimately connected with medical uses, as bio-based materials have been sought for use in numerous applications. Using just one established example, metal parts have been used in knee replacement operations for decades. Routinely, some metal part is inserted as an alloy cap for both the shinbone as well as the thighbone. If cartilage has been damaged, a choice of plastics can be inserted to replace that which is damaged. But now, surgeons have a wider suite of materials at their disposal. We discuss this below.

https://doi.org/10.1515/9783112205822-013

13.2.1 Bone mimics

A significant amount of work has gone into the production of bone mimics. The use of titanium as a repalcement for bones that have been diseased, damaged, or injured goes back generations. It has been used extensively for knee and hip replacements because the metal does not react with the bone and tissue around it, is relatively low in density, and functions without degradation for decades. But stainless steel and cobalt–chromium alloys have also been used.

Because natural bone is a complex material, it is difficult to replicate it, especially in high enough quality that the human body does not reject the material as foreign. Hydroxyapatite has been used in several studies of bone graft material – $Ca_{10}(PO_4)_6$ $(OH)_2$. A significant amount of effort is placed on producing hydroxyapatite with pore sizes comparable to that in natural bone because this increases overall strength and appears to reduce any negative interactions. The basic synthesis for nonnaturally oc-curing hydroxyapatite is shown in Figure 13.1.

$$6\,H_3PO_4 + 10\,Ca(OH)_2 \longrightarrow Ca_{10}(PO_4)_6(OH)_{2(s)} + 18\,H_2O$$

Figure 13.1: Hydroxyapatite synthesis.

A great deal of the hydroxyapatite that is grown is done so in a clean enough environment that the end material finds use in dental implants. But its applications are often dictated by the technique in which it is formed, which can result in different-shaped and different-sized crystlas.

Calcium phosphates have also found use in bone regeneration. They can be used alone or with materials like hydroxyapatite to stimulate bone growth. As well, calcium sulfate has found use in this application in large part because it is biodegradable in the body, as bone regenration occurs.

We have also mentioned metal foams in Chapter 8 and point out here that because of their low density, now also, they are becoming a material of choice for biomedical implants for both bones and dental prostheses [6].

13.2.2 Contact lenses

The earliest contact lenses were made from precisely shaped glass, although research and development into the idea meant that what could be called a second generation was made of the plastic polymethylmethacrylate (PMMA). Even these early plastic lenses required time for the wearer to become accustomed to them. In short, wearing them could be painful at first. What are called soft contact lenses were first intro-duced in the 1960s. As the name implies, they were far more flexible than previous lenses and are porous enough that they are gas permeable. Improvements in the feel of contact lenses and their gas permeability are continuing areas of research.

Currently, there is a class of contact lenses that are termed extended wear. Many of these contact lenses are made of silicone hydrogels, which enable them to absorb water so that the lens does not become dry against the eye and allows oxygen to permeate the lens and interact with the eye in a healthy manner, as shown in Figure 13.2 [5]. These lenses can now be worn for weeks at a time if needed then changed for a new pair.

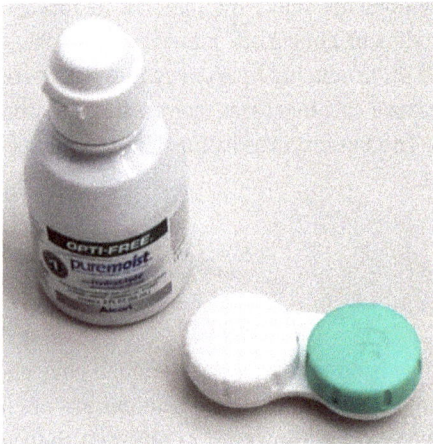

Figure 13.2: Contact lenses.

13.2.3 Body parts

Use of titanium for joint replacement has a decade-long history, but is pre-dated by the use of titanium in dentistry in the 1940s. Properties that make titanium extremely useful in terms of body repair include:
1. Fracture resistance
2. High strength
3. Nonreactivity with living tissue
4. Long-term, good mechanical performance

But these properties are important in virtually any type of body part replacement. A table showing all types of body part replacements that are possible would be extremely lengthy [6]. We have included in Table 13.1 the body parts that are most commonly replaced and listed the type of material often used for that part. Also, we will note that in every case possible, current research involves finding some animal source for the replacement material, in large part because it can be made biocompatible with the patient's body. As well, many animal-based replacement parts and tissue can be absorbed into the body over time.

Table 13.1: Biomaterials and body parts.

Body part	Material	Strengths	Comment
Arm/leg	Plastics, rubber point of connection	Strength of prosthetic	Arm, electrical impulses can be triggered by arm movement. Made as composites, so the prosthetic responds like a natural arm/leg
Artery	Nickel-titanium, various plastics	Lasts for full life of the patient	Can be made to elute a drug over time
Bone	Calcium phosphate, calcium sulfate, hydroxyapatite	Injectable into the patient	
Dental implants	Tetragonal zirconia (ZrO_2) and yttria (Y_2O_3)	Highly durable	With 3% yttria, abbreviated Y-TZP, a ceramic
Ear tissue	Can use patient's living cells	Biocompatibility	
Eye	Acrylic	Provides normal appearance	Ocular implants that detect light are now possible
Heart valve	Caged ball valve, silicone ball in steel/titanium cage	Long-lasting	Current research, use of porcine tissue
Hip joint	Aluminum oxide (Al_2O_3) ceramic	Inert in the body	Natural bone can grow intoporous aluminum oxide during healing
Jaw	Titanium	Stronger than bone section it is replacing, inert	
Testicle	Rubber or saline-filled plastic	Provides normal appearance	

13.2.4 Hernia mesh

We will discuss hernia meshes because hernia surgery, especially abdominal wall hernia reconstruction and repair, is one of the most common surgeries performed in the United States each year. More than a million such surgical procedures have been performed annually for the past several years. A wide variety of activities cause hernias (such as repeated heavy lifting), but hernias can actually occur in infants as well as adults.

To use just one example of a biomaterial which can serve to illustrate a greater whole, meshes made from traditional plastics are currently used extensively in a variety of types of hernia repairs. Such meshes are usually polypropylene, but can be polyester or polytetrafluoroethylene as well. All are used because they have been found to be generally unreactive with human tissue over a long period of time, certainly within the span of time that is required for a patient to heal after a hernia surgery. But all are used to hold a sewn patch of tissue in place while it heals, something which may take a few months. Yet such hernia meshes will routinely be implanted in patients for as long as 40 or 50 years since there is no generally accepted medical procedure to explant them after some set period of time.

Prior to the use of any sort of nonreactive plastic hernia mesh, surgeons operated without such materials and had to advise patients not to exert themselves while they healed. Hernia meshes of any sort thus increased the rate of successful surgeries. Yet plastic meshes were not really made to be in a human body for decades. The question then arises as to what might be a better solution.

Hernia surgeries often involve a surgeon deciding on the spot whether or not a patient requires a mesh after suturing so that the sutures do not rip when a person undergoes some violent, involuntary action, which can be as simple but disruptive as a sneeze. What can now be called traditional hernia meshes: the just-mentioned polypropylene, polytetrafluoroethylene, or polyester materials are nondegradable within the body. Figures 13.3 and 13.4 show meshes from two different manufacturers. Figure 13.3 is noted as a polypropylene material.

Figure 13.3: Polypropylene hernia mesh.

Biologically based hernia meshes are now being fielded, with different companies using different material. At least one firm is using tissue that is essentially linings from pigs, while another is using similar materials obtained from sheep. Additionally, other firms are experimenting with producing synthetic proteins that can be taken up by the body over the long term, when a mesh is no longer needed.

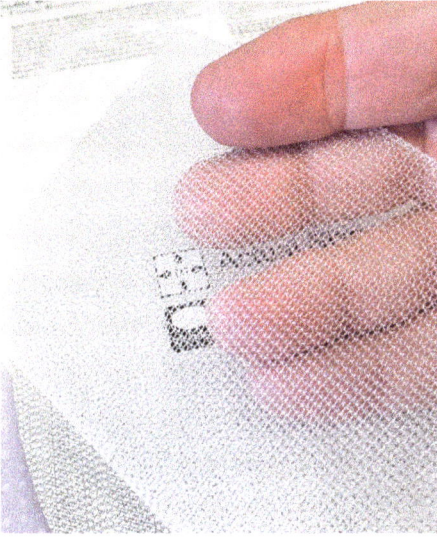

Figure 13.4: Hernia mesh.

In the past decade, as mentioned, more than one company has experimented with producing some type of hernia mesh made from other living tissues. In listing them as categories, examples now include:

1. Bovine pericardium – a layer of membrane tissue that covers the heart muscle. Cows are the source.
2. Porcine intestine submucosa – connective tissue derived from the intestines of pigs
3. Porcine skin – skin layers derived from pigs.
4. Human acellular dermal matrix – the term "acellular" means cells are removed from human tissue samples; yet the support structure for the tissue remains.

These are becoming more common with the passing of time and as word of their effectiveness spreads among the medical community. These materials eventually are absorbed or degraded into the body without complications, unlike the three just-mentioned plastics.

13.2.5 Skin grafting and repair

Before the advent of biomaterials directed to skin regeneration and repair, medical professionals at times used samples of a patient's own skin to repair some spot that had suffered severe trauma or simply relied on bandages and a sterile environment for regrowth to occur. This was not possible in cases in which a patient suffered severe burns or other trauma over large parts of their body. But the recognition that skin can be transferred from one part of a body to another actually goes back several hundred years.

Perhaps, one of the oldest, most colorful, and yet-documented stories of skin grafting is that of nasal reconstruction to counter one of the effects of syphilis. One of the effects of untreated syphilis is the degenration of the nasal bridge and overall nose. First documented in the 1500s, the patient had a flap of skin from their arm surgically attached to the degraded bridge of the nose while the other end of the flap remained attached to the arm to supply blood. The patient then had the arm strapped to their head with a leather harness until such time as blood vessels formed from the flap to the face, oftentimes 2 weeks. A second surgery was then performed to cut the skin flap from the arm. While this seems horribly primitive today, it was an effective means of using a biomaterial to perform a type of reconstructive facial surgery [7].

As with the body part replacements we have tabulated and discussed, animal sources for skin replacement have been used. Called xenografts because it is a foreign material, such tissue is harvested from a variety of animals, although porcine skin grafts have become common as the practice has spread in the medical community. They are often used as temporary cover for burned skin.

13.2.6 Silk

The natural material silk has been harvested from the cocoons of silkworms – the bombyx mori, or silk moth – for centuries. For most of this time, it has been used to make luxury garments. The modern means of producing natural silk is through what is called sericulture. The worms are cultivated on mulberry plants, where they eat the leaves, then cocoon themselves for their metamorphosis. The cocoons are harvested and boiled to produce the best silk, a protein; and throughout history, this has been a valuable material for clothing, as a trade item between East and West. In some of the lands where silk is produced, the people who do such work are devoutly Buddhist and do not wish to kill any living thing, including cocooned worms. They prefer to let the moth chew its way out of the cocoon. While such silk can still be used, the hole in the cocoon means the silk cannot be extruded as a single, long fiber, and thus the end product is considered inferior.

While the protein that makes up silk is complex, a repeat unit of it often is made of four amino acids: serien, glycine, alanine, and glycine. Figure 13.5 shows this.

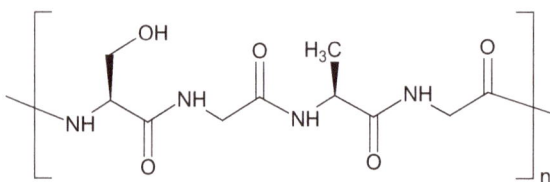

Figure 13.5: Repeat structure of silk.

Silk sutures for wound repair and surgical thread is a material of choice because it does not react with tissues, and bioerodes in no more than approximately 2 years.

13.3 Mycelium-based end-user items

We have mentioned in Chapter 1 that wood is one of the traditional materials which has found use in an almost limitless number of ways throughout history, from making small objects and furniture, to large constuctions, such as buildings and ships. Very recently, efforts have been made to use mycelium – a material that is the root structure of mushrooms – to produce furniture, leather, and other items [8, 9]. What had been nothing more than science fiction a decade ago has become something of a niche in the furniture and the leather markets. Mycelium materials are grown into the desired shape and can then the resulting object can be used – such as a piece of furniture – or further shaped as desired. One firm that has become a leader in this area, Mylo, says concerning this material:

> Made from mycelium, the underground root-like system of fungi, Mylo™ is a bio-based leather alternative that is soft, supple, and less harmful to the environment [9].

The means by which mycelium is grown into desired shapes can be proprietary to different companies, but it is not hard to visualize growing any plant-based material into a certain shape using molds while it grows. This is the technique by which automobile parts, such as the soft paneling in doors, can be made from shaped mycelium.

Other advantages of using mycelium in applications where wood or plastics have been used in the past include:
1. Mycelium adds fire resistance to building materials, as it does not burn quickly.
2. Imitates the feel and look of leather – this appeals to those trying to live a vegan lifestyle.
3. Can be used with types of brick, as a type of natural bonding material.
4. Has found use as insulation – as a replacement for petroleum-based plastic materials.
5. What are called "fungal biorefineries" can produce mycelium to order, as construction materials [10].

As mycelium emerges and claims a larger market share where it competes with other materials, emphasis is placed on its environmental friendliness as well as its long-term sustainability.

13.4 Biofuels

The term "biofuel" simply means an existing form of fuel that is produced from a re-newable source, often plant matter and not from crude oil. Since materials science and materials chemistry tend to focus on solid matter, there is disagreement as to whether or not biofuels qualify as some class of biomaterial. We will treat this class of materials here because it is a biologically based series of sources for what has always been petroleum-based products.

Bioethanol, bio-butanol, and biodiesel are all chemically equivalent to their coun-terparts which are refined from crude oil. But their sources are different. Fueling sta-tions now often show separate prices for gasoline and any biofuels they sell, such as 85% ethanol, which is sold as E-85. Table 13.2 gives some basic information about each. Figure 13.6 shows one such filling station, with prices for both traditional gaso-line blends and E-85, or 85% ethanol, fuel.

Table 13.2: Biofuel sources and production.

Biofuel	Sources	Annual production	Annual production of non-bioversion
Ethanol	Corn, sugar beet, sugarcane	355.3 M gallons	15B gallons
Butanol	Simple sugars, corn, grains	*	
Diesel	Palm oil, rapeseed oil, sunflower oil, animal fats, restaurant yellow grease	2.2B gallons	69,7B gallonss

*Alternative Fuels Data Center [11], although production numbers for biobutanol do not appear to be published.

13.5 Bioplastics

Earlier, in Chapter 7, our discussion of polymers had a focus on the six polymers that are produced in such large amounts that they have their own RICs. In the past three decades, a growing amount of research has been directed at what are called bioplas-tics. It is fair to say that virtually every plastic made in large amounts has been exam-ined for some medical use, as they tend to be flexible and are often inert. But some have been more widely embraced than others.

There appears to be a twin set of aims in research directed at bioplastics. The first is to produce some plastic from a renewable source material. As an example, if polyethylene is routinely sourced from ethylene which in turn comes from crude oil, producing it from ethylene that is ultimately sourced from some plant would create a process in which the production of this plastic is now biologically based.

Figure 13.6: Biofuel advertised at a gas station.

The second aim in the production of bioplastics is to produce plastics that biodegrade relatively quickly, in a matter of months or years, as opposed to decades or centuries. As an example, polylactic acid (PLA) is a biodegradable polymer that can be used for trash bags, landfilled, and allowed to degrade.

Perhaps, obviously, the ultimate bioplastics are those which have a renewable source, and which are readily biodegradable. We list here those bioplastics that have seen the greatest use and that continue to see high levels of use, both medically and in society in general.

13.5.1 Polylactic acid

The polymer PLA can be produced from the fermentation of corn, sugarcane, and other plant matter. The Lewis structure of its repeat unit is shown in Figure 13.7. The material finds extensive use in food packaging material since it is biodegradable. As well, it continues to find increasing uses in a variety of medical implants because it can be absorbed by the body over time.

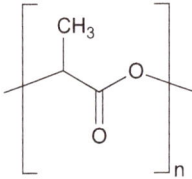

Figure 13.7: Unit structure of polylactic acid.

13.5.2 Polyethylene

Common, starchy materials such as sugarcane or corn can be fermented to produce ethanol, which can then be dehydrated to produce ethyelene. The simplified chemistry is shown in Figure 13.8. This ethylene is then polymerized as any other ethylene would be resulting in polyethylene.

Note that polyethylene is not biodegradable, whether it is produced from bio-sourced materials or not. But polyethylene is produced in such large quantities each year that substitutes for the source ethylene are constantly being pursued.

$$C_6H_{12}O_{6(s)} \longrightarrow 2\,CO_{2(g)} + 2C_2H_5OH_{(l)}$$

then

$$C_2H_5OH \longrightarrow H_2O_{(g)} + C_2H_{4(g)}$$

Figure 13.8: Bioethylene production.

13.5.3 Polyhydroxybutyrate

Sometimes more precisely called poly-3-hydroxybutyrate (PHB), this material can be biologically synthesized from several microorganisms. Industrially, it continues to be produced from materials sourced from crude oil. Its repeat structure is shown in Figure 13.9. Note that, like PLA, it falls into the category of plastics called polyesters. Numerous polyesters have been used in some biomedical application, often because they are both flexible and compatible with tissue in the body.

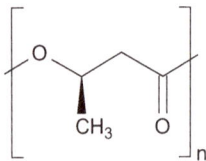

Figure 13.9: PHB repeat structure.

13.5.4 Cellulose

Cellulose is used in a wide variety of applications. Medically, it has been used in controlled drug delivery systems, dressings for wounds, and even in some blood purification systems. It is nontoxic and compatible with tissues in the human body. It is generally produced enzymatically, by cellulose synthase. The Lewis structure of the cellulose repeat unit is shown in Figure 13.10.

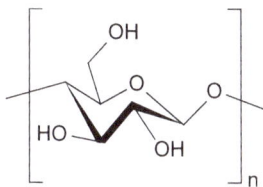

Figure 13.10: Cellulose structure.

13.5.5 Polycaprolactone

Polycaprolactone (often abbreviated PCL) has found use in a variety of surgical implants and has also proven to be useful in drug delivery systems. The synthesis of it involves the ring opening of caprolactone. Since this is a single reactant reaction, the product can often be produced in high purity. But the source material for caprolactone remains based in crude oil. Figure 13.11 shows the polymerization that produces PCL.

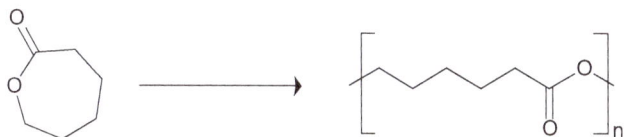

Figure 13.11: Polycaprolactone production.

As we have mentioned, there are many more plastics that have found some medical use. There is no overarching theory as to why a specific plastic is used in one application. Rather, this is often a matter of trial-and-error, starting with some animal model such as mice or rats.

13.6 Biomaterials for nonmedical applications

Biomaterials have been widely used for medical applications, as the discussion in this chapter makes clear. Advances in the use of such materials have unfortunately sometimes been driven forward by modern wars and the medical attention required to save lives in them. But because of the increase in the availability of such materials, from wars or through other means, biomaterials are also finding uses in nonmedical applications. Examples we have already seen include:

1. Silk – for example: clothing, parachutes, upholstery, wall "paper."
2. Mycelium – for furniture and structural parts, and for imitation leather.
3. Plastics – for an enormous variety of user end items.
4. Metals – again, for an enormous variety of uses and user end items.

Undoubtedly, the uses of biomaterials will continue to expand. The reasons will routinely be the efficiency of a specific biomaterial for an application as well as a potential cost savings in some established procedure or in the manufacture of some product.

References

[1] Society for Biomaterials (SFB). Website. (Accessed 16 June 2025, as: https://biomaterials.org).
[2] Canadian Biomaterials Society. Website. (Accessed 16 June 2025, as: https://biomaterials.ca).
[3] European Society for Biomaterials. Website. (Accessed 16 June 2025, as: https://www.esbiomaterials.eu).
[4] The Australasian Society for Biomaterials and Tissue Engineering (ASBTE). Website. (Accessed 16 June 2025, as: https://www.asbte.org).
[5] All About Vision: What are contacts made of? Website. (Accessed 16 June 2025, as: https://www.allaboutvision.com/eyewear/contact-lenses).
[6] A. Rabiei. Recent developments and the future of bone mimicking: Materials for use in biomedical implants, *Expert Review of Medical Devices*, 2010, 7(6): 727–729.
[7] The Renaissance Nose Job: The Official Website of Author William Eamon. Website. (Accessed 16 June 2025, as: https://williameamon.com/the-renaissance-nose-job).
[8] The Future Is Made Out of Mushrooms. Website. (Accessed 6 May 2022, as: https://www.nationalgeographic.com/science/article/mushroom-fungi-furniture-video-spd).
[9] Bolt Technology – Meet Mylo ™ (Accessed 16 June 2025 as: https://boltthreads.com/technology/mylo/).
[10] M. Jones, A. Mautner, S. Luenco, A. Bismarck, S. John. Engineered mycelium composite construction materials from fungal biorefineries: A critical review, *Materials & Design*, Feb 2020, 187: 1–16. (https://doi.org/10.1016/j.matdes.2019.108397).
[11] Alternative Fuels Data Center. Website. (Accessed 16 June 2025, as: https://afdc.energy.gov).

Chapter 14
Ceramics

14.1 Introduction, history, basic compositions

Working natural clays is one of the oldest human activities that can be considered a form of chemistry and chemical transformation, along with refining and forging metals such as copper, tin, and iron or with making fermented beverages. The only chemical transformation we might claim came before these was the taming of fire in some long-forgotten past. Silicate materials dug directly from the earth were mixed in some careful, controlled way with water and possibly some additive or additives, shaped into some object that is useful, then solidified into that shape by placing the object in a kiln or open fire for some amount of time. This intense heating process drives the water from the clay, with the result being a rigid, strong, but oftentimes brittle object. The many clay objects and fragments of objects found in archaeological dig sites in our modern era are a testament to how durable clay objects are since some of these are several thousands of years old.

The terms clay and ceramic are often used interchangeably by the general public, but can be defined more specifically. Clay is generally hydrous aluminosilicates, often mixed with other minerals, which is highly deformable when wet, but which is hard, rigid, and often brittle when dried. A ceramic is more precisely defined as a crystalline or glassy nonmetallic, nonorganic material that is hard and often brittle after it has been heated. Since some ceramics are not silicates, it is fair to make the claim that clays are a subset of the larger category of ceramics. The colors of natural clays are because of various minerals that are in the overall material, such as iron oxides. Materials such as silicon carbide (SiC) are an example of a type of ceramic that contains no silicates.

Additionally, we generally consider clay or ceramic objects – even ancient ones – to be those made with the help of a wheel on which a lump of clay is spun and shaped. Many cultures did discover and use the wheel, although others worked in and made usable objects from clays, and appear not to have done so. Figure 14.1 shows a modern reproduction of a famous find from the Cahokia Mounds near modern-day St. Louis, Missouri, USA. The bowl is shaped with a bird's head, but the object was modeled entirely by hand, without the aid of a pottery wheel. It appears that the people of this vanished, pre-Columbian culture worked clays found near where they lived, but may never have discovered the idea of a pottery wheel or at least not used it extensively. Figure 14.2 shows an example of ancient clay item that did require a wheel. The photo is of an oil lamp of the type mentioned in ancient texts, such as the Bible. To create this, it was first spun on a wheel, then pinched at one side to create a channel, and it was fired to harden it. To use it, oil was poured in it, and the pinched channel held the wick.

https://doi.org/10.1515/9783112205822-014

Figure 14.1: Modern reproduction of clay bowl from the Cahokia Mounds.

Figure 14.2: Ancient oil lamp made of clay.

Very often, clay is the material associated with the production of brick walls and brick buildings. The surface of such walls that is in view can be of any sort of finish, smooth or rough. Figure 14.3, for example, shows a small portion of a brick wall of a house. The coloring is because some amount of one or more iron oxides were part of the formulation.

It is incredibly obvious to people that in brick walls, bricks are routinely laid in patterns in which they are staggered, not stacked – so often that it seems painfully simple to mention it. But the reason is that the binder that joins one row of bricks to another, as in Figure 14.3, is not strong enough to withstand the extremes of stress that a wall might undergo. This back-and-forth stacking uses the strength of the bricks to reinforce the strength of the wall.

The field of clays and ceramics is so large that an American Ceramics Society and several other national or international organizations devoted to ceramics exist [1–11], all of which represent the manufacturers and companies which produce such materials, and many of the end user items made of these.

While the structure of ceramics can differ wildly, based on what components are in a particular formula, the repeat unit of (SiO_2) coming together as linked tetrahedra

Figure 14.3: Common bricks that make up a wall.

in many ceramics creates crystalline patterns, even if they are only microcrystalline. Figure 14.4 shows a silica tetrahedron.

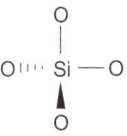

Figure 14.4: Silica tetrahedron.

If the silicon atom is considered the center of a tetrahedron, linked tetrahedra routinely share two of the oxygen atoms which form the corner of each tetrahedron. This is why the repeat unit of silica is routinely listed as SiO_2 and not SiO_4. Figure 14.5 shows common ways in which tetrahedra can be linked in crystalline ceramic materials, with the central point of what appears to be each triangle being the silicon atom. The figure is not inclusive of all possible means of forming repeating patterns, but rather some very common ones.

14.2 Ceramic production and properties

Some unified theory of ceramic production has never existed, nor does one exist today, despite considerable research into the field. Incredibly, the entire field is one of systematic trial and error, whether for decorative ceramic works, such as finished pottery vessels, or for precise applications, such as catalytic converter supports in automobiles or semiconductors. Common steps in ceramic production tend to include the following:

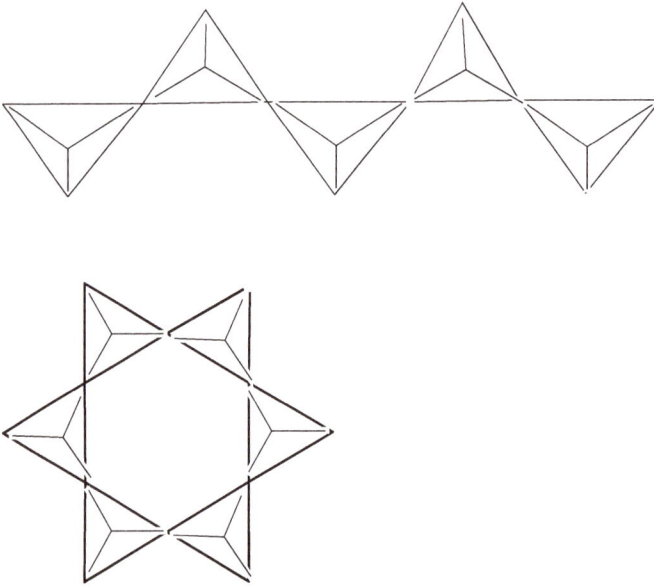

Figure 14.5: Examples of linked tetrahedra.

1. Beneficiation: A form of purification of the starting material since naturally gathered or mined clay, silica, and alumina can have significant amounts of impurities in it. Beneficiation concentrates the desired materials – such as silica – that will be used in making some finished ceramic material.
2. General mixing: This involves clay, additive oxides, and water, always in precise ratios once a formula has been determined.
3. Shaping: The desired object must be put into a form. This can be done by traditional shaping by hand or on a wheel, or by pouring slip, a mostly liquid suspension of particles in water, or by machine-forming.
4. Heating: Virtually all ceramics undergo some high temperature process for the final formation of the end product. This is almost always a kiln firing. Table 14.1 shows several temperature zones for firing of different ceramic materials.
5. Glazing: This step is not used in all applications. For example, ceramic brick for building walls is often not glazed. Ceramic tiling that is in some way decorative is almost always glazed.
6. A second firing: After glazing with what is often a wet material, the once-fired object is again glazed to give it color and a particular finish.

Interestingly, there are several different categories into which ceramics can be divided. One of the major set of divisions is shown here, in Table 14.1. This gives some idea of the categories, and the criteria by which categories are determined. But it is worth keeping in mind that other possible breakdowns can be constructed as well.

Table 14.1: Categories of ceramic materials.

Ceramic category	Subdivisions	Examples	Firing temp. (°C)	Comments
Bone China		Plates, tableware	1260	Bone ash, 50%; kaolin, 25%; and often feldspar 25%
Earthernware	Glazed	Pottery		Generally: clay, 25%, quartz, 35%; kaolin, 25%; feldspar, 15%
	Unglazed	Terra cotta, tiles, roof tiles	550	Can absorb water
Nonoxide	Borides	SiB_3, SiB_6 coating for space shuttle tiles		B_4C used in body armor
	Carbides	SiC		Widely used as abrasives, used in body armor
	Nitrides	Industrial abrasives		
	Silicides	$MoSi_2$, in heating elements	>2,000	
Porcelain		Traditional tableware		First developed and produced in China
Stoneware		Chemically resistant components	1,000–1,200	
	Alumina	Spark plugs, artificial joints	1,560–1,700	
	Ceria	CeO_2, specialized electrodes		
	Silica	SiO_2		Extremely wide variety of uses: fiber optics
	Zirconia	ZrO_2 in dental implants	1,170–2,370	Dental implants because of biocompatibility

Table 14.1 shows that there are similarities in composition between the different forms of ceramics, and yet, there are ceramic categories that can be entirely different in composition than what might be expected. The nonoxide ceramics are a firm example of this; yet, they are still classified as ceramic materials.

Of note, Table 14.1 mentions alumina, zirconia, and ceria. All of these can be structurally similar to silica, in that the metal atom can occupy the center of a tetrahedron which in turn is part of a much larger series of tetrahedra. But uses are not determined based on their repeating unit structure. Rather, uses are based on known physical characteristics. For example, zirconia is used as a grinding material precisely because it is harder than alumina.

14.3 Major uses of ceramics

Ceramics have long been used in a wide variety of applications because they can be shaped into virtually any form imaginable. One of the oldest examples are building bricks [11]. As well, they are used because once they are fired, they are extremely hard and resilient. The multitude of ceramics that have been excavated worldwide from archaeological sites is proof of the age-long durability as well as the hardness of ceramics. At the same time, as is well known, ceramics tend to be brittle and thus break rather than deform. A literal mountain of broken pottery near Rome that exists because each of the containers were broken and discarded after their cargoes were delivered definitely testifies to this. Archaeologists have named this site "Monte Testaccio" or "Broken Pot Mountain."

An enormous array of ceramic items are well known not only to scientists but also to the general public, with several being used as consumer end items such as many types of tableware. Several have been noted in Table 14.1. Some applications are not as widely known, or perhaps we should say not as widely considered. These include the following sections of Chapter 14.3, which is a nonexhaustive list.

14.3.1 Roofing tile

One of the most distinctive style of roofs in the western world is what is called the Spanish tile roof. Made of terra cotta, and having a distinctive red color, such roof material is resistant to the changes in weather and extremely durable. Some roofs made of this material have lasted for centuries. Figure 14.6 shows one example.

14.3.2 Floor tile

Floor tiling has been used for millennia, often as a sign of status and wealth when placed in homes. The ceramic material from which it is made has routinely been the product of local manufacturing. Although as modern nation-states emerged, some types of tiles, often glazed with vibrant colors, were transported for long distances to become part of the homes of the wealthy. In these cases, such ceramic material was not designed for heavy use, but rather for durable yet decorative use. Figure 14.7 shows an example of such ceramic tiling. Note that the surface upon which people walk is unadorned, while the vertical surfaces are attractive and designed to be eye-catching.

Figure 14.6: Spanish tile roof.

Figure 14.7: Functional and decorative floor tiling.

The vertical, facing tiles shown in Figure 14.7, the photo of steps in a building entranceway, are decorative and thus do not need to be able to support heavy weight or tolerate harsh conditions. But notice on the third step down, the sixth tile from the left, a single piece was put in place with the floral tile design upside down in relation to all the others. This is an example of a largely bygone practice in which the craftsman "signed" his work. An intentional change in the pattern like this was a means by which the person working the project could make some claim to it, to take credit for the craftsmanship. In Figure 14.7, we have indicated where this tile is with an arrow.

14.3.3 Wall tiling and facing, brick work

These just-mentioned applications are often considered in terms of a very simple characteristic: they look attractive. Even roofing tile and brick work, which are very often thought of in terms of how little water permeation they allow, have been used in applications in a wide variety of buildings where they are seen, and where their visual appeal is important.

Floor tiling and wall facings have for centuries been signs of wealth and status for the people in whose homes they were placed. Roman villas sometimes had mosaics made of ceramic tiling, some of which still survive today, and much of which is both detailed and attractive. The formulas of such materials were never standardized.

Today, the manufacture of bricks is an enormous industry. Companies such as Acme Brick in Texas have revenue that is close to one billion dollars annually – this being only one of many companies that produce bricks.

There are enough such firms that trade organizations exist to promote the manufacture and sale of bricks and tiling [7, 8, 11]. As might be expected, they advocate for the use of brick as opposed to any of several other construction materials.

14.3.4 Tableware and cookware

Many people immediately think of tableware or cookware when the term "ceramic" is used. A very old term for such items, "China," is still used for much decorative table ware and is an indicator of its origin. In the latter half of the eighteenth century, two companies emerged which produced decorative, attractive ceramic items for the wealthy of Europe: Wedgewood and Meissen. In England, Josiah Wedgewood produced his now-famous Wedgewood blue [12], an example of which is seen in Figure 14.8.

At roughly the same time, in what is now Germany, Johann Friedrich Böttger produced a red porcelain ceramic that would become as famous in continental Europe as Wedgewood china did in England [13]. Figure 14.9 shows an example of a Meissen porcelain medal that honors Böttger and that adheres to the original formulation from local materials. Curiously, Böttger found that replacing the local red clay with a white kaolin produced a white "China" after firing. This is considered the first European china, and because Böttger lived under a polite house arrest for many years, dictated by the Elector of Saxony, the recipe remained a tightly guarded secret.

The goal of both companies, at least at their outset, was to find some formula for "china" and thus break the hold that companies in the Far East had on this material – and in the process reap a tremendous profit.

Figure 14.8: Examples of Wedgewood items.

Figure 14.9: Meissen porcelain medallion.

As might be expected, although the formulas of both of these ceramics was closely guarded and remain proprietary today – and although the red color in Meissen porcelain is believed to come from an iron oxide in the formula – both formulas have been mimicked by numerous other companies. But also of note is that each of these two companies has expanded far beyond their original, signature looks, and produce large amounts of different types of ceramics today [12, 13].

14.3.5 Kiln and oven linings

The use of ceramics for the linings of ovens is another application that goes back millennia, and such materials are often called firebrick. More recently, it is called refractory. These names do not lead us to any one specific formula, and indeed numerous formulae have been used over time – usually a mixture of clay, alumina, magnesia, and silica, at times in a proprietary formula. Table 14.2 shows several of the possibilities, but is not an exhaustive list.

Table 14.2: Firebrick components.

Name	Component possibilities	Formula
Alumina		Al_2O_3
	Bauxite	$Al(OH)_3$ (impure)
	Diaspore	$AlO(OH)$
	Kyanite	Al_2SiO_5
Silica		SiO_2
	Quartzite	SiO_2 (powder)
	Sand	SiO_2 (granular)
Magnesia		MgO
	Dolomite	$CaMg(CO_3)_2$
	Forsterite	Mg_2SiO_4
	Magnesite	$MgCO_3$
	Olivine	$(Mg,Fe)_2SiO_4$

The overarching requirements for firebrick are that the material be able to withstand high heat, and that it has a relatively low thermal conductivity. Firebrick must be able to retain its shape, and not fracture, with repeated changes in temperature as an oven or kiln is heated and cooled. Note that, in Table 14.2, the possibilities for materials that can be used in a formulation are wide. Once again, the choice is often dependent upon the use or application.

14.3.6 Ballistic body armor inserts

As ceramics have continued to develop, and new formulas that combine light weight with extreme hardness have been found, ceramics have found a growing market in modern body armor. Used by law enforcement and militaries throughout the world, plates can be inserted into nylon garments – the garment often being Kevlar – and provide the wearer with enhanced personal protection. Such plates, often called trauma plates, can be made of metals or plastics, such as titanium, or of ultrahigh molecular weight polyethylene. But the combination of lighter weight and resistance to deformation makes ceramic materials a good choice as well. The ceramics that have been found to be most effective are carbides, such as SiC or boron carbide (B_4C). These plates may break upon impact, but can be replaced in the garment in which they are encased.

14.3.7 Armored vehicle plating and reinforcement

These final two categories of ceramic use are quite recent and use ceramic materials in a nontraditional manner. In both, the ceramic plate or piece is utilized because it is both hard and relatively light – which is important for a soldier wearing such plates – but also takes advantage of the brittleness of ceramics. When hit by some ordnance, the ceramic plating usually breaks or shatters. But it can be replaced, and the protective coating of the person or vehicle is again restored.

The first use of modern ceramic armor appears to be in United States helicopters during the Viet Nam War. The lighter weight of ceramics when compared to metal alloys made them useful as deck plating in helicopters where weight is always of concern. Carbides and borides were the most common formulas, specifically: SiC, B_4C, and sometimes titanium boride.

One thoroughly modern use of protective ceramic plating is the outer skin of the NASA space shuttles. The ceramic tiling is able to withstand the heat of re-entry to the atmosphere at the high speed with which any spacecraft is traveling. These tiles are silicates, sometimes called HRSI, for high-temperature reusable surface insulation. They are capable of withstanding temperatures up to 1,250 °C – meaning that much like the just-mentioned kiln linings they have extremely low thermal transmission.

References

[1]   United States Advanced Ceramics Association. Website. (Accessed 16 June 2025, as: https://advanced ceramics.org).
[2]   The American Ceramics Society. Website. (Accessed 16 June 2025, as: https://ceramics.org).
[3]   Association of American Ceramic Component Manufacturers: AACCM. Website. (Accessed 16 June 2025, as: https://aaccm.org).
[4]   British Ceramic Confederation. Website. (Accessed 16 June 2025, as: https://www.ceramics-uk.org).
[5]   English Ceramic Circle – British Ceramics. Website. (Accessed 16 June 2025, as: https://www.english ceramiccircle.org.uk).
[6]   British Ceramic Research Association. Website. (Accessed 16 June 2025, as: https://www.luci deon.com).
[7]   Cerame-Unie, The European Ceramic Industry Association. Website. (Accessed 16 June 2025, as: https://cerameunie.eu).
[8]   European Ceramic Tile Manufacturers' Federation (CET). Website. (Accessed 16 June 2025, as: htps://uia.org).
[9]   The Australian Ceramics Association. Website. (Accessed 16 June 2025, as: https://australianceram ics.com).
[10] Ceramics Southern Africa. Website. (Accessed 16 June 2025, as: https://ceramicssa.co.za).
[11] Brick Industry Association. Website. (Accessed 16 June 2025, as: https://www.gobrick.com).
[12] Wedgewood. Website. (Accessed 16 June 2025, https://www.wedgewood.com/en-us).
[13] Meissen Porcelain. (Accessed 16 June 2025, as: https://www.meissen.com).

Chapter 15
Glasses

15.1 Introduction and history

The very first glass materials believed to have been known in history are the placer obsidian pieces that early man picked up and used as some sort of cutting tool or weapon. Archaeologists continue to debate when the first examples of this naturally occurring volcanic glass were used. But sites around Mount Arteni in modern Armenia have been called "a Stone Age weapons factory" [1].

With this in mind, as defined in the widest possible way, glass can be any amorphous, solid, nonmetallic material which exists as a solid at ambient temperature, is brittle when solid, but that can be transformed into a molten, soft yet viscous liquid material. Also, glass can be transformed between the solid and liquid state repeatedly. The general public tends to consider glass in less technical terms, thinking of it as in some way transparent, or as a hard yet brittle material that is definitely not at all metallic. The general public tends to think of glass as the materials for windows or as a material for making dinnerware and drinking glasses, as shown in Figure 15.1. Utilizing this in terms of a broad definition, glass could be considered a polymer, based on repeating units of silicon oxides. This chapter however will discuss traditional types of glass, meaning the end material produced from silica and sometimes other inorganic materials.

Figure 15.1: Common drinking glasses.

The production of specific types of glass has an ancient history, almost as old as the production of ceramic pottery and refined metals. Glass objects have been something of a luxury item for much of that time. Its manufacture appears to be somewhat younger than that used in the kilns that produced the first pottery, but close in age to the

https://doi.org/10.1515/9783112205822-015

forge technology that was used to produce the first metals, such as copper, bronze, and iron. It appears then that the production of ancient types of glass may have sprung from these other two technologies, although this is still debated among archaeologists and historians. What appears to be certain is that the development of the blow pipe, to increase the heat of fires used to make glass, appears in the archaeological record in the first century BC, in the area at the eastern end of the Mediterranean Sea.

As mentioned, the truly earliest glass was not produced so much as found. It is the naturally black volcanic glass known as obsidian. It was used as cutting tools, and items such as spear points, and not fashioned beyond this.

Man-made glass has traditionally been produced by fusing silica sand with lime and soda, and on occasion other ingredients, to produce a hard material that has been desired for its transparency.

When glass was first used in windows – roughly, during the Middle Ages – the production of it in sheets was limited because of the size of furnaces that could melt glass, and the use of blow tubes by glassblowers. This is why the stained glass windows of the cathedrals of Europe are made of relatively small panes of glass that have

Figure 15.2: Church window.

been connected with some metal solder, often lead. Figure 15.2 shows a modern church window, which is made of much larger sheets of glass, but which still uses metal partitions, in large part to imitate the look of church windows made in what can be called the traditional manner.

15.2 Basic compositions

The three basic materials required for the production of glass will appear to be very close to those which we discussed in Chapter 14, ceramics. They are silicon dioxide, calcium oxide (aka, lime), and soda ash. In virtually every glass-making operation, the silicon dioxide – colloquially called "sand" – must be of high purity. All three of these components are tracked by the USGS Mineral Commodity Summaries annually [2], although it does not track production of glass independently. Soda ash is used largely in the production of glass; but perhaps obviously, sand finds much greater use in the production of cement and concrete, which we will discuss in Chapter 16. Calcium oxide also has another major use, the production of steel.

At the simplest, a formula for glass requires silicon dioxide (silica) and sodium oxide. It can be represented broadly as shown in Figure 15.3, which tries to show the possible variation in stoichiometry.

$$SiO_2 + (1/4 - 3/4)Na_2O \rightarrow Na_2O \cdot SiO_2$$

Figure 15.3: Glass production chemistry.

A variety of other metal oxides besides sodium oxide can be used in glass formation, but sodium oxide remains the most common today. Adding such materials when glass is being formed is routinely done so to achieve some desired outcome in terms of the properties of the resulting glass. As might be expected, a great deal of research has gone into making the types of glass that are resistant to deformation – glasses that do not break when something impacts it. Table 15.1 shows a number of different types of glass, but is not an all-inclusive list. The applications column to the right gives some uses of a particular type of glass. The numbers in the 12 columns from SiO_2 to GeO_2 are percentages of the material that is in that particular glass formula. Columns with an "X" in them are so designated because a formula is proprietary to a particular firm.

Soda–lime glass is the type made in the largest amount by far, Table 15.2 illustrates the different possibilities that exist for different types of glass. Just over 90% of all glass continues to be soda–lime. Overall, the volume of glass produced annually and the myriad uses for it is such that several trade organizations have been established to promote the sale and use of various types of glass [3–9].

Table 15.1: Common glass formulas.

Type	SiO$_2$	Na$_2$O	K$_2$O	Al$_2$O$_3$	CaO	MgO	B$_2$O$_3$	PbO	ZnO	BaO	GeO$_2$	Applications
Aluminosilicate	57			16	10	7	4			6		Reinforced plastics, fiber glass
Alkali-barium silicate								X		X		Television screens, X-ray absorption
Fused silica	100											High-temperature applications
Lead oxide	59	2	12	0.5				25	1.5			Low-temperature applications
Oxide	90										10	Fiber optics: High clarity, wide-angle lenses.
Sodium borosilicate	81	4.5		2			12					Pyrex: High-temperature, cooking materials
Soda–lime–silica	72	14.2		0.6	10	2.5						Windows, drinking glasses. Largest use.

15.3 Glass blowing and the Pilkington process

Despite glass having been produced for various uses for more than 2,000 years, there exist only two ways to produce it on a large scale. Glass blowing has ancient origins stretching all the way back to the time of the Roman Empire. At that time, and for centuries after, glass was made into bottles, cups, and other relatively small containers.

Traditional glassblowing routinely undergoes several broad steps including the following:

1. Batch house: The required materials are mixed at this point. They will be fed into the furnace to be melted.
2. Furnace, aka, the hot end: At this point, raw glass is melted, and the molten material often formed into some object, sometimes using molds. Glass that is to be blown can be formed into a semimolten ball at the end of the glassblower's blowpipe.
3. Annealing: Slow cooling of glass objects is extremely important to ensure that they are not easily breakable. Objects are fed into or simply placed into an annealing oven. This ensures controlled and slow cooling. Depending on the size and thickness of the object, annealing operations may go on for days.
4. Cold end: This represents the last step of the production of glass objects, and by necessity includes an inspection of the object for any visible defects. At this point, finishes such as details or logos may be attached and annealed to the object.

This process is extremely energy intensive, as is the next one we will discuss, usually requiring furnace temperatures of 1,750 °C. Either oil or natural gas is routinely used as the fuel source to generate the heats necessary to melt the sand and other raw materials in the glass batches.

The shift from being able to produce flat glass for windows from a relatively small-sized pane to the large sheets of glass common today began in the nineteenth century, but took a revolutionary step forward much more recently, in 1959, with what is called the Pilkington Process. This process, now often called the float glass process, is originally named after Sir Alastair Pilkington, who along with Kenneth Bickerstaff patented a revolutionary means by which molten glass could be formed in perfectly flat sheets by being laid on a bath of molten elemental tin. The process takes advantage of the immiscibility of molten glass in molten tin. Molten glass is laid onto a bath of molten tin that is often 60 m long, and as it travels down the length of the tin bath, it cools from a molten state to a solid state [10]. This ensures that the resulting glass is perfectly smooth, and also ensures it is highly durable, as the cooling process is a slow, controlled one.

By the 1960s, the Pilkington process had replaced essentially all others in the production of flat glass or plate glass. The production of bottles, drinking glasses, and other glass objects however still utilize some type of lathe in many cases, which turns the glass object that is being made.

Figure 15.4 shows the basic schematic of a float glass operation.

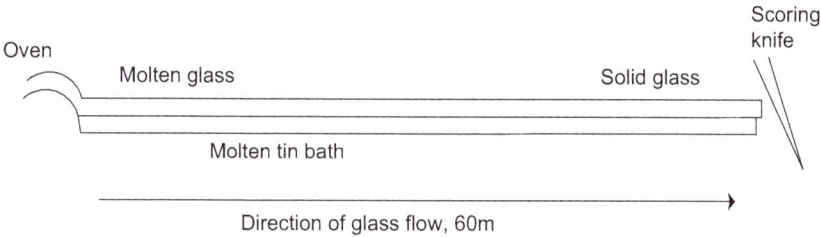

Figure 15.4: Basic schematic of a float glass operation.

Note that the traditional ingredients – sodium carbonate (aka, soda ash), sodium sulfate (aka, salt cake), sand (silicon dioxide), limestone, dolomite, and sometimes other materials used as coloring agents – are first mixed in batches, then introduced to a furnace. This increases the temperature of the material to 1,500 °C. This now molten glass of whatever composition is then added atop the molten tin bed – the tin bath – in a controlled manner, so that the glass weighs enough to flatten itself out on the tin not cause ripples in the molten metal. As the glass batch, now a ribbon, moves down the tin, it cools to approximately 600 °C, where it can be removed.

After removal from the tin bath, flat glass must still be cooled slowly. This annealing process prevents stress and cracking in the finished, flattened glass. At this point, it can be cut into lengths to be used for windows.

15.4 Optical cabling

Optical cabling has found wide use in the telecommunications industry. The cabling is more expensive to produce than traditional copper wiring, but can pass more information through a single cabling through total internal reflection of any signal. This is possible when the glass fiber is extremely pure, and thus production methods require both high precision in the mixing of starting materials, and an extremely clean environment in which the process is undertaken.

The production of optical cabling involves spinning glass into extremely fine, hair-like fibers. Also, the core, the center portion of the cabling which transmits the signal, is a component made of glass. Surrounding layers sheath the core and protect it from damage, such as kinking or cracking.

To create the material from which an optical fiber is made, a preform, essentially a glass tube, is first made. This glass tube is heated on a lathe and rotated to ensure particles that can be deposited on it evenly. A gaseous material such as silicon tetrachloride (SiCl$_4$) or germanium tetrachloride (GeCl$_4$) is injected into the middle of the preform while it is on a lathe, and at the elevated temperature of approximately 2,000 °C, this reacts with the effluent of a hydrogen–oxygen flame, depositing pure silicon dioxide on the interior wall of the glass tube. A representation is shown in Figure 15.5.

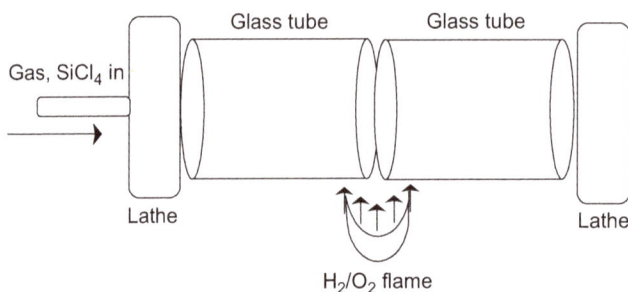

Figure 15.5: Preform positioned for silicon dioxide deposition.

The preform and deposited glass is then heated to the point at which the preform collapses, making the deposited inner core into a single strand.

The collapsed preform/core is then attached to a vertical drying tower, and the core is then extruded into progressively finer wiring, with its vertical arrangement taking advantage of gravity, and being assisted by a ball of glass being attached to the end of the fiber. This pulls the material until the resulting fiber is approximately the thickness of a human hair.

15.5 Bullet-proof glass

The old term "bullet-proof glass" has been replaced within the security industry with the more appropriate term "bullet-resistant" glass. It is also called transparent armor or ballistic glass. As all these names imply, the use of such material is to deflect and stop projectiles. It is used in transports for VIPs, in the windows of high-value properties such as jewelry stores, and in various military vehicles.

Interestingly, while bullet-resistant glass is made with a significant amount of glass, it often has a polymer as an inner layer. Various plastics can be used and ethylene vinyl acetate has been proven to be one of the most effective inner layers. Layers of glass are then bonded to the inner plastic. This provides three layers, which can be made stronger by laminating even more layers on these. Often, a thin layer of a plastic such as polycarbonate is the final layer on the inside (the side sometimes termed as the safe side). This prevents any spalling of particles on the person or persons on that side of the glass.

Despite the strength of such materials, the term "bullet-proof" is less used now than in the past, simply because it is recognized that larger and higher caliber projectiles have been made, some of which will ultimately overcome the strength of even this type of glass.

15.6 Additives and optical properties

An entire spectrum of colors can be imparted to glass by adding various materials to the molten glass batch. Scientists are often familiar with brown or red glass that is used when making compounds that are light sensitive – the dark color helps protect and keep the solution within stable. Various shades of red scientific glassware are made for the same purpose. Figure 15.6 shows an example.

Figure 15.6: Colored scientific glassware.

One example of a type of colored glass that has become famous over time is the addition of colloidal gold or gold chloride to molten glass to produce a ruby red color, simply for its attractiveness. Sometimes called cranberry glass because of its appearance, this color has routinely been produced only in small amounts because of the price of gold, and for many years was associated with the manufactories of Venice. Figure 15.7

Figure 15.7: Red glass made with gold additive.

shows a gold-infused glass flower vase with this color, made in Venice in the traditional manner.

Table 15.2 gives some examples of the colors that can be formed in glass using different additives. Since there is no overarching theory about color formation – it is much more a trial-and-error system – the table is arranged by the old mnemonic, "Roy G Biv" for the visible spectrum. And perhaps obviously, the concentration of additive can determine the depth of color – red versus pink, for example.

In most cases in Table 15.1, we have not included percentages of a particular additive. This is because small changes in the amount of additive can shift a color. For example, small amounts of an additive that produces a red glass may produce a pink or yellow shade. Some companies keep their formulas as trade secrets.

The additives listed in Table 15.1 can be incorporated into glass mixtures when they are molten, semimolten, or before initial melting, to provide the desired color. An old but established term for such types is pot metal glass, when metal oxides are used as additives, and when they are added while the glass is molten. As well, addi-

Table 15.2: Additives for glass colors.

Color	Additive	% Added	Comments
Red	Gold metal	0.001%	Traditionally expensive
Pink	Gold metal	<0.001%	–
Red	Copper metal	–	Laminated onto clear glass
Red	Selenium	–	Sometimes as a combination: Cd, Se, S
Red	Uranium	–	Added to leaded glass
Orange	Tin	–	Added to leaded glass
Yellow	Silver nitrate	–	Surface coated
Yellow	Sulfur, iron salts	–	Variable concentration, from yellow → brown.
Green	Uranium	<2%	Yellow in lower concentrations
Green	Copper oxide	–	Variable concentrations, from green → blue
Blue	Nickel	–	Variable concentration, from blue → black
Blue	Cobalt	–	Variable concentration, from blue → purple
Violet	Manganese	–	–
Opaque white	SnO_2, or As_xO_y	–	Can be through surface treatments.

tives can be applied to a surface and then re-heated. This fuses a colored layer to a clear glass base. The term flashed glass is used to describe this technique and result.

Stained glass windows are perhaps the example of colored glasses best known to the public. To achieve a specific effect, often numerous small panes of a colored glass are combined to make a greater whole that is visually different from any one color. As an architectural technique, stained glass appears to have begun at the Augsburg Cathedral in Germany. Since then it has spread to churches and cathedrals worldwide, and still is a use for colored glass.

15.7 Uses of types of glass

To list all the uses of all types of glass would be a lengthy proposition. Here, we have quoted the Glass Alliance Europe website, which goes far beyond the idea of window, bottles, and jars that are probably what the general public thinks of when considering uses for glass. The website states:

"Glass is used in the following non-exhaustive list of products:
- Packaging (jars for food, bottles for drinks, flacon for cosmetics and pharmaceuticals)
- Tableware (drinking glasses, plate, cups, bowls)
- Housing and buildings (windows, facades, conservatory, insulation, reinforcement structures)
- Interior design and furniture (mirrors, partitions, balustrades, tables, shelves, lighting)

- Appliances and Electronics (oven doors, cook top, TV, computer screens, smartphones)
- Automotive and transport (windscreens, backlights, light weight but reinforced structural components of cars, aircrafts, ships, etc.)
- Medical technology, biotechnology, life science engineering, optical glass
- Radiation protection from X-Rays (radiology) and gamma-rays (nuclear)
- Fibre optic cables (phones, TV, computer: to carry information)
- Renewable energy (solar-energy glass, wind turbines)" [9]

Even this list most likely omits smaller, niche uses for specialty glasses and glass materials.

15.8 Glass recycling

Glass is one of the most recycled and reused materials in the world, and has a history of recycling and reuse that is older than almost any other material (some metals have seen small-scale recycling for over a century). Glass objects such as bottles can be crushed, melted, and reformed, or can simply be sterilized and reused. Additionally, glass can be crushed back to powder-sized particles and used in applications not usually thought of, such as road or levee fill, to stabilize an area where traffic or water can erode a surface.

In traditional recycling operations for glass bottles and jars, they are placed or dropped into large, collecting containers. When full, such containers are then given or sold to companies which can remelt the batch. From this, they can produce new glass containers. Cullet is the term used for glass that has been recycled in this manner. But glass bottles can be recycled without breaking and remelting the first bottles. In such instances, used bottles are returned to the company that used them, sanitized in boiling water, and then simply reused.

Glass recycling can be the policy of a national government, or in the cases of large nations, can be policy or law dictated by state or regional governments.

References

[1]   "Paleolithic weapons factory was a rich source of obsidian tools from 1.4 million years ago." M. Miller, *Ancient Origins*. Website. https://www.ancient-origins.net, (Accessed 16 June 2025).
[2]   U.S. Geological Survey, Mineral Commodity Summaries 2024. Downloadable.
[3]   Glass Association of North America. Website. (Accessed 16 June 2025 as: https://www.glass.org).
[4]   National Glass Association. Website. (Accessed 16 June 2025, as: https://www.glass.org/).

[5]   Glass Manufacturing Industry Council. Website. (Accessed 16 June 2025, as: https://gmic.org/.

[6]   Glass Build America. Website. (Accessed 16 June 2025, as: https://www.glassbuildamerica.com/).

[7]   British Glass. Website. (Accessed 16 June 2025, as: https://www.britglass.org.uk/).

[8]   Bundesverband Glasindustrie. Website. (Accessed 16 June 2025, as: https://www.bvglas.de/en/).

[9]   Glass Alliance Europe. Website. (Accessed 16 June 2025, as: https://glassallianceeurope.eu/ s).

[10]   Pilkington. Website. (Accessed 16 June 2025, as: https://www.pilkington.com).

Chapter 16
Cement and concrete

16.1 Introduction and history

The chemistry involved in making cement is not as old as that of brewing beer or wine, or of extracting metals from ores, but it still has millennia-old ancient history. Some cement formula was used by the peoples of ancient China, the Roman Empire, and ancient Southwest Asia, as has been verified through the analysis of materials at archaeological dig sites. One amazing accomplishment of engineers in the time of the early Roman Empire was the production of a type of cement that set underwater. Some structures made using it still stand today.

The more modern history of cement formulations only goes back to the nineteenth century. It is believed that Joseph Aspdin was one of the first people to approach the production of cement in a systematic and scientific manner. His mixtures of clay and limestone resulted in some of the first modern cements. He was the first to patent what is now known as Portland cement – made from limestone.

The development of new types of cement continues, simply because the modern world relies heavily on it, and uses this class of materials in new ways and under progressively more extreme conditions. The dams, buildings, and roadways of the modern world all require huge amounts of cement, and require that it be extremely long lasting. The international production of Portland cement is shown in Figure 16.1 as a fraction of the 4.081 billion of metric tons produced in 2020. While the share produced by China is over half of the total, it is worth noting that China is building at an extremely rapid pace right now; and it is unclear how long such a pace can be maintained.

What the general population considers cement invariably involves one or more trucks with the distinctive rotating drum, from which the wet material is poured (a material more properly called concrete). An example is shown in Figure 16.2. But there is much more to cement than just this.

A discussion of cement can be divided quite simply into cement which can set underwater – called hydraulic cement – and that which cannot – called nonhydraulic cement. Note that here the word "concrete" indicates cement-based mixtures that also include materials such as sand, stone, or any other aggregate material.

16.2 Geographic sources

The raw materials for all common types of cement are common in most parts of the world. The United States Geological Survey tracks cement production in thousands of metric tons in its annual Mineral Commodity Summaries [1] and indicates that lime-

https://doi.org/10.1515/9783112205822-016

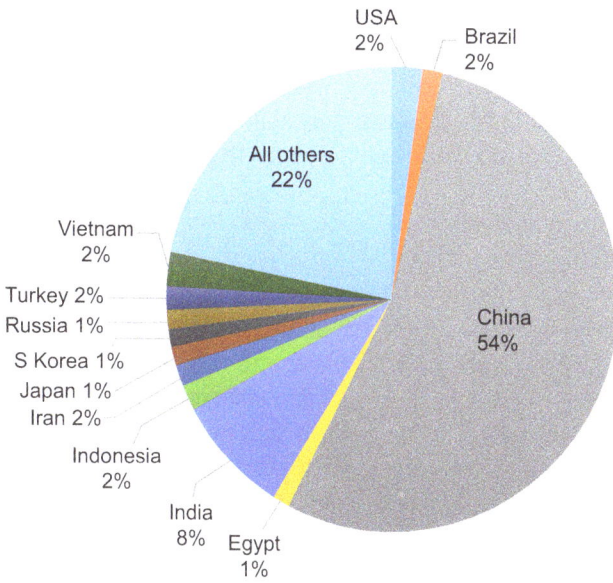

Figure 16.1: Cement production by nation.

Figure 16.2: Cement truck and road repair.

stone – one of the chief components – is abundant in most areas. This being said, the United States does import cement in some instances, from Canada and Mexico, as well as from South Korea and China. Because cement is so widely used, it is not surprising that several national and international trade associations that are dedicated to advocating for the production, sales, and uses of cement and concrete exist [2–10].

16.3 Cement formulation chemistry

Because there are several types of cement, there are different categories into which they can be placed. As just mentioned, the most simple one involves its ability to set under water – hydraulic cement – or not set under water – nonhydraulic cement. All hydraulic cements must have one or more components in their formulation that form what are called sparingly soluble hydrates. For nonhydraulic cements to set, some amount of carbon dioxide is needed. This can be as simple as carbon dioxide from the local atmosphere, from the air.

16.3.1 Hydraulic cement: reaction chemistry

Portland cement has a 200-year history. Joseph Aspdin, the first to patent a formula for it in 1824, gave it the name because he felt it looked like the stone building material quarried from the Isle of Portland, off the south coast of England. Mixed calcium–aluminum oxides are a major component of Portland cement. Routinely made through what is termed the dry method, the following steps are required in its manufacture:
1. Raw material crushing, or quarrying and crushing: This step brings limestone ($CaCO_3$) and clay to particle sizes of 6–7 cm.
2. Blending: In this step, both silicon dioxide and calcium oxide (SiO_2 and CaO) – fly ash – as well as iron ore are added to the crushed material.
3. Kiln processing: In this step, a kiln that is angled up from the horizontal, and that is at approximately 1,500 °C, rotates as the material passes through it. This step removes any gaseous material and in the process forms clinker, pieces that are roughly 2 cm in size.
4. Cooling: The clinker is brought to lower temperatures.
5. Grinding: This again brings the final product to smaller particle size, and is coupled with limestone and gypsum ($CaSO_4 \cdot 2H_2O$) being added.

The major difference between this dry method and what is termed the wet method is that in the latter method, the material is processed as liquid slurry [2].

It is difficult to represent the formation of cement in terms of stoichiometric chemical equations, but Figures 16.3 and 16.4 are attempts to do so. It should be noted

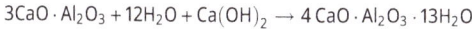

$$3CaO \cdot Al_2O_3 + 12H_2O + Ca(OH)_2 \rightarrow 4CaO \cdot Al_2O_3 \cdot 13H_2O$$

then

$$(8+x)H_2O + (3CaO \cdot Al_2O_3)_2 \rightarrow 2CaO \cdot Al_2O_3 \cdot 8H_2O + 4CaO \cdot Al_2O_3 \cdot xH_2O$$

then

$$4CaO \cdot Al_2O_3 \cdot Fe_2O_3 + 7H_2O \rightarrow 3CaO \cdot Al_2O_3 \cdot 6H_2O + CaO \cdot Fe_2O_3 \cdot H_2O$$

Figure 16.3: Hydraulic cement setting.

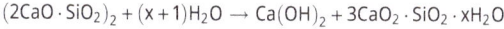

$$(2CaO \cdot SiO_2)_2 + (x+1)H_2O \rightarrow Ca(OH)_2 + 3CaO_2 \cdot SiO_2 \cdot xH_2O$$

then

$$(3CaO \cdot SiO_2)_2 + (x+3)H_2O \rightarrow 3Ca(OH)_2 + 3CaO_2 \cdot SiO_2 \cdot xH_2O$$

Figure 16.4: Hydraulic cement hardening.

that to form a solid material, the end result must involve the calcium–aluminum material first setting then hardening.

The reactants as well as products of the reactions in Figure 16.3 may at first glance appear complex or imprecise, but each is an example of the formation of some type of metal oxide hydrates. For the hardening in hydraulic cement, the reactions shown in Figure 16.4 show similar product formation.

Sparingly soluble hydrates are the products formed in these two reactions, essentially solid solutions. These reactions are dependent on water. As a consequence, they can still occur under water.

The formulation of a particular batch of cement can change depending on the specific need, but all are extremely durable and long lasting as materials. For example, the materials used to construct the now ruined piers at the ancient city of Caesarea when it was part of the Roman Empire, off the coast of modern Israel today, were made of materials quite like modern mixtures.

16.3.2 Chemistry: nonhydraulic cements

The reaction of a starting material with carbon dioxide is the force which drives the chemical formation of a nonhydraulic cement mixture. The use of carbon dioxide in this case can be seen in the reactions shown in Figure 16.5.

The above three reactions in Figure 16.5 appear to end right where they began, but there is more to them than that which can be shown in the reactions alone. First, calcium oxide must be produced in a purified form. This requires high heat – approximately 850 °C. The pure calcium oxide is then reacted with water to form calcium hydroxide, still called slaked lime in industry. After this, carbon dioxide hardens the slaked lime as calcium carbonate.

$$CaCO_{3(s)} \rightarrow CaO + CO_2$$

and

$$CaO + H_2O \rightarrow Ca(OH)_2$$

and

$$CO_{2(g)} + Ca(OH)_2 \rightarrow CaCO_{3(s)} + H_2O$$

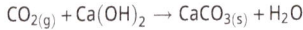

Figure 16.5: Nonhydraulic cement formation.

Calcium hydroxide can also react with what is called pozzolan – a mixture of silicon-based or aluminum-based oxides also containing iron oxides – to form what the American Concrete Institute terms "compounds having cementatious properties" [2]. The pozzolan must be in finely divided form to optimize the reaction with the calcium hydroxide.

16.3.3 Cement: other types

There are several other formulations of cement besides Portland cement and those we have just discussed. The following may have roots that are historical, but for a large-scale use they represent relatively recent advances in cement technology and materials science. All have found increased use in the past few decades.

– Blast furnace slag cement
As the name implies, this type of cement utilizes the slag that is tapped off of blast furnaces. Slag is a mixture of the silicates found in iron ores, which are separated from the reduced iron and removed from a blast furnace as a molten material. Slag is less dense than molten iron, and thus is always tapped from a port located higher in the blast furnace than the port which taps the iron.

For much of the past, slag was simply discarded as a waste material, but more recently it has been used in various industrial situations. There is even a National Slag Association as well as a Slag Cement Association that advocates for the use of slag here and in other applications (such as road bed material) [12, 13].

In using blast furnace slag, after it is removed from the furnace, it is quickly quenched, then ground into fine powder after being dried. Solidifying the slag quickly means it becomes a material that can be described as glassy both in appearance and at a molecular level, and that importantly can be pulverized and made usable. As high as 90% slag can be used in the formulation of this kind of cement.

– Fly ash cement
Once again, as the name implies, fly ash cement utilizes the fly ash – also known as coal ash or flue ash – which is recovered from coal-fired power plants. As with blast furnace slag, the composition of fly ash is a mixture of silicate materials that are sim-

ply a combustion by-product in the burning of coal. Up to 40% fly ash can be used in a cement mix of this type.

– Rapid hardening cement

Cements that are considered to be rapid hardening are those in which the clinker has approximately 5% of gypsum ($CaSO_4 \cdot 2 \cdot H_2O$) added to it after the gypsum itself is finely ground. In this case, the term finely ground means that it is determined to have particles that give it a high surface area, generally 500–600 m^2/kg. Most other cements have particle size that equate to roughly half this area.

While there are several applications for rapid hardening cement, the air forces of developed nations tend to use it in what are called rapid runway repair kits. Should a runway be bombed by an enemy force, it can be repaired and used again in the shortest possible time, often less than one day.

– Energetically modified cement

Energetically modified cement (EMC) is made from pozzolan-like materials, and gets the term because of a relatively recent development called high-energy ball milling. Through this type of milling, the materials – which are still volcanic ash, blast furnace slag, and Portland cement – are ground to an extremely fine powder. Since this is done using high-energy ball milling, the overall input of energy is lower than in traditional cement production methods.

A ball mill is a rotating chamber with balls within it that constantly impact the material being ground down to some smaller particle size. In the production of EMC, the balls are able to reduce the material to a much finer powder than older, traditional means of milling and grinding.

At the EMC Cement website, the organization touts how little carbon dioxide is produced as a by-product of EMC manufacture, even claiming it is a "big, fat zero" [14] While this is obviously motivated by a section of industry trying to emphasize how environmentally friendly it is, in an industry that has traditionally not been considered to be so, it does bring attention to the continued efforts to lower the coproduction of this greenhouse gas when cements are used.

– Autoclaved aerated concrete (AAC).

Another type of cement-based material, based on the way it is formed, autoclaved aerated concrete (AAC) finds considerable use as a building material. It is produced using finely ground quartz sand as well as calcined gypsum, lime, and water. Powdered aluminum can be utilized as an additive, in small amounts ≈0.05%. It is first mixed, to a point where hydrogen bubbles form. The autoclave step takes place after the material has been cut into desired shapes, and this enhanced pressure lasts for at least 12 h at elevated temperature, roughly 190–200 °C. At this temperature and autoclave pressure, hydrated calcium silicates form, resulting in a material with exceptionally high strength.

AAC is suitable for use in interior construction applications, and its low density (ca. 20% of the weight of other concretes) is its major advantage. First patented in the 1920s, research and work continues on this form of cement, with recent patents still being issued [11].

16.4 Cement and concrete: major uses

The overwhelming use of cement is as a binder in a myriad of materials used in construction. The building material concrete is the mixture of cement with stone and sand. The uses of cement and concrete are numerous, but major uses include the following, and is not an exhaustive listing.
1. Construction
   a. Roads
   b. Housing
   c. Nonresidential buildings
   d. Dams – some of the largest man-made structures ever produced.
   e. Sewer piping
   f. Marine construction
2. Decorative
   a. Residential and commercial building patios
   b. Driveways
   c. Pool decking

Figure 16.6 shows a view of a raised municipal parking deck that is made of concrete. The structural supports include rebar – reinforcing iron bars within the concrete – but beyond this structural enhancement, the deck is essentially concrete. Such parking structures are common in cities and other areas where land is at a premium.

16.5 Cement recycling

Cement and concrete are not usually considered recyclable materials in the same sense as glass, plastic, paper, metal, and wood. This is because the cement or concrete applications utilized in a residential building are most often walls and other supporting structures and are designed to last for decades or even centuries. Yet cement and concrete that have been used in some way – for example, in the production of roads – can be ground into small pieces and in some manner reused. This may not be in another batch of cement for continued road construction. Rather, it may be as the base material for a side-of-road sound deadening berm. Especially if such a use is less expensive overall than the use of other such materials, the incentive exists to reuse cement.

Figure 16.6: Concrete raised parking deck.

References

[1]   U.S. Geological Survey, Mineral Commodity Summaries 2024. Downloadable.
[2]   American Concrete Institute. Website. (Accessed 14 June 2025, as: https://www.concrete.org/).
[3]   Concrete Society of Southern Africa. Website. (Accessed 14 June 2025, as: https://concretesocietysa.org.za).
[4]   Concrete Institute of Australia. Website. (Accessed 14 June 2025, as: https://concreteinstitute.com.au).
[5]   Cement Association of Canada. Website. (Accessed 14 June 2025, as: https://cement.ca/).
[6]   CEMBUREAU. European Cement Association. Website. (Accessed 14 June 2025, as: https://www.cembureau.eu/).
[7]   Association of Cementitious Materials Producers. Website. (Accessed 14 June 2025, as: https://acmpsa.org.za/).
[8]   Concrete Masonry Association of California and Nevada. Website. (Accessed 14 June 2025, as: https://cmacn.org).

[9]   Mineral Products Association (MPA) Cement. Website. (Accessed 14 June 2025, as: https://cement.mineralproducts.org).

[10]   Cement Industry Federation. Website. (Accessed 14 June 2025, as: http://cement.org.au).

[11]   Autoclaved aerated concrete structure components. C.M. Hunt., US Patent 8,720,133 B1, 2014.

[12]   National Slag Association. Website. (Accessed 14 June 2025, as: https://nationalslag.org).

[13]   Slag Cement Association. Website. (Accessed 14 June 2025, as: https://slagcement.org).

[14]   EMC Cement. Website. (Accessed 14 June 2025, as: https://www.emccement.com).

Chapter 17
Lightweight materials

17.1 Introduction

In previous chapters, we have discussed many types of materials that have been made for specific applications exactly because they are lightweight because their density is low. Metals such as aluminum and magnesium, as well as their alloys, are often used in this regard. Additionally, many plastics have found specific applications precisely because they are light weight when compared with any other materials that could be used in the same application.

Curiously, some of what might be considered the "oldest" of the lightweight materials are natural materials such as spruce wood and ash wood. Both were used to construct the earliest airplanes. Since the engine was made of metal and could not conveniently be lightened with the existing technology of the early twentieth century, woods that were both strong and light were used for parts of the frames. Likewise, cork was used in many cases in which some low density material was needed.

Today, there are many needs for lightweight materials, and some organizations devoted to the idea [1, 2]. What is called the light-weighting of automobiles becomes a career specialty for some people working in one of the auto manufacturing companies. Plastic parts are often used in lieu of metal ones where the end weight is lighter, and safety is not compromised. As well, using light-weight materials is critically important in the aerospace industry, whether it is related to the production of passenger aircraft, or the much smaller field of producing unmanned and manned spacecraft and satellites.

17.2 Basic classifications

Lightweight materials can be divided into broad categories, including metals, ceramics, and plastics. It is perhaps obvious that these subjects have been covered separately in different chapters in this book. Here, we will compare and contrast their macroscopic properties and also discuss their uses.

Table 17.1 is a non-exhaustive list of materials that have found uses and applications because they represent a lightweight option.

A number of materials have been produced that are of extremely low density because of the means by which they are made. In Table 17.1, we have already noted the idea of metal foams. In such extremely lightweight materials, usually air or some other gas is injected into the material at some stage of its production, lowering the density of the resulting material or object. Perhaps, the most common example of this kind of material synthesis is Styrofoam® – more properly, extruded polystyrene (PS) foam. Figure 17.1

https://doi.org/10.1515/9783112205822-017

Table 17.1: Lightweight material categories.

Material	Formulas	Applications/uses	Comments
Plastics	See Chapter 7	Automotive, construction	Often to replace metal parts
Mycelium	Biological	Construction materials	As an environmentally friendly option
Metals	Al, Mg, alloys thereof	Aerospace industry	Can have Zr as an alloy component
Metal foams	Some metal with air or nitrogen blown through	Automotive	Energy absorption, shock absorption
Ceramics	Silicon carbide	Spacecraft	Provides thermal protection as well

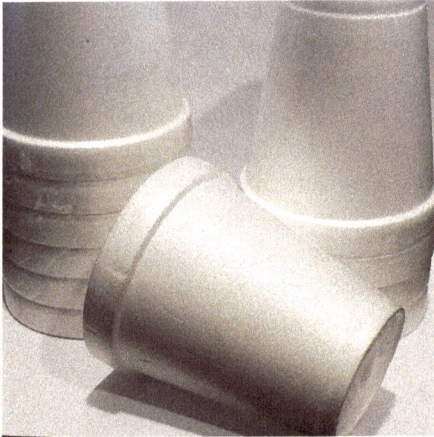

Figure 17.1: Styrofoam® cups.

shows common Styrofoam® cups, but there are numerous other items made from it as well.

The general public probably considers Styrofoam cups and plates as an example of widely known low density plastic. But this material is also used as packing material extensively both because it is light in weight and because it performs well in absorbing shocks that can occur when items are being transported. This actually speaks to what might be called the expanded idea of lightweighting any item: there is usually some other property besides low density that is also advantageous in the end product.

Table 17.2 gives examples of several extremely low density materials. Some are very common, and others, such as aerographite, are recent inventions, having only been first reported in the past 10 years [3]. Still others, such as metal foams made using aluminum, have found use in automobile parts and production [4]. They are both light in weight and still sturdy enough to perform well.

Table 17.2: Foams and aerated materials.

Material	Density (g/cc)	Uses
Aerographite	0.0002	First reported in 2012
Metal foams	0.0009	Heat exchangers, flame suppression
Aerogel	0.001	Usually a silica base
Styrofoam	0.075	Packaging, disposable dinnerware

17.2.1 Metal materials

In Chapter 8, we discussed aluminum, magnesium, and their alloys and noted that such alloys routinely find use in the aerospace industry, including in the construction of space-launched vehicles. Table 17.3 shows a tiny number of the many alloys of aluminum and magnesium that have found some use. Not all such alloys are used simply because they are of low density. As well, we have listed some other metals that are of relatively low density, but that have found niche uses at most. Scandium is an example of a low density metal with several uses, all of them rather small scale.

Table 17.3: List of metal densities.

Metal	Density (g/cc)	Comments
Aluminum	2.712	
Aluminum alloy 3003	2.730	Widely used in sheet metal, food containers, cookware
Duralumin	2.790	Contains 4% Cu. Can contain other element in small amounts.
Lithium	0.534	Used increasingly in EV auto batteries
Beryllium	1.85	Numerous niche uses
Magnesium	1.738	
Magnesium AZ31B	1.770	Contains Al and Zn; used in aircraft bodies
Magnesium AZ91D	1.810	Used in die casting
Scandium	2.99	Very few uses

Lithium is the least dense of metals and thus is of interest in several low-density applications. It is the choice of all automobile manufacturers who are launching or planning to launch an advanced electric vehicle – for use in the batteries. We have included lithium in Table 17.3, but will note that lithium is actually a relatively scarce element, largely because there is no nucleosynthetic pathway that ends in a particular state that is energetically favorable at some isotope of lithium. Thus, it is not widely distributed throughout the planet. The USGS Mineral Commodity Summaries track lithium annually [5] but do not disclose the amount produced domestically in the United States – Nevada being the one area with a currently active mine.

Figure 17.2 shows the breakdown of lithium that is mined globally. Note that while Australia has the largest percentage at this time, the entire total of all production is only 184,300 metric tons. When compared to metals like aluminum, this total is quite small.

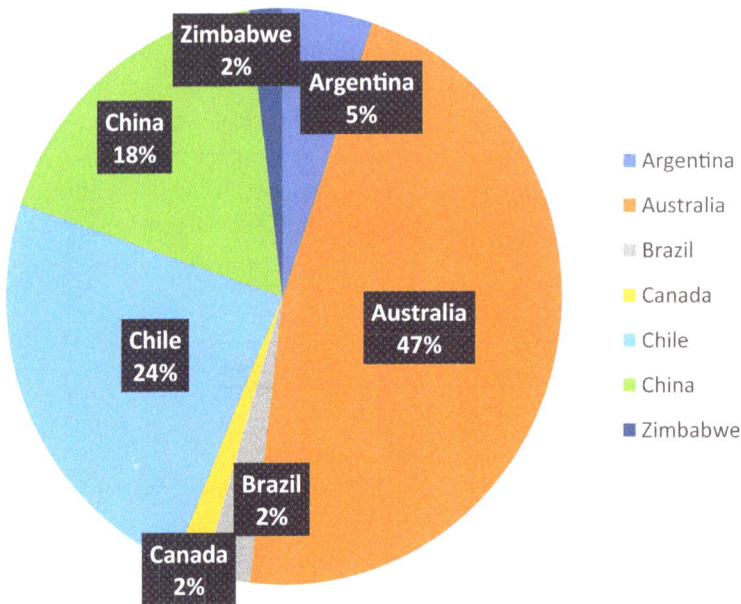

Figure 17.2: Global lithium production.

17.2.2 Ceramic materials

Not all ceramic materials are considered lightweight, but in comparison to some metals – in instances where both can serve the same application – ceramics are often considered to be the material of lighter weight.

It is difficult to compose a table of densities for ceramic materials because the formulations can change even from one batch of material to another. Still, examples of what can be called lower density ceramics continue to emerge.

One example of where ceramics are used in place of metals that has grown in the past decade is what is called advanced modular armor protection. Used in various armored vehicles worldwide, including in the German Leopard tanks of the Bundeswehr, such ceramic armor does shatter with sufficient impact, but can be replaced. Interestingly, it has also found use in a limited way in some military helicopters – where low overall weight is important – although the chemical formulation has not been made public.

We will also mention concrete at this point, although it may seem amusing to consider this a lightweight material. We make note of it because the American Society of Civil Engineers continues to sponsor a concrete canoe competition among colleges and universities each year. The event has become so well established that there is even a website devoted to it [6]. The main thrust of the competition is to create a canoe that can seat four people that is made of some form of concrete. All ideas about making such canoes less dense are encouraged among contestant teams.

17.3 Polymers and plastics

We have seen a brief table of densities for metals. Since they can be purified to elemental materials, it is relatively easy to determine their densities. When making similar comparisons among plastics, there is more variability to densities. This is because there are differences in how plastics are made from company to company, and even possible differences in how one batch is made when compared with another. Table 17.4 presents several plastics that are produced on a large scale and their densities.

Table 17.4: Plastics, densities.

Plastic	Abbreviation	RIC	Density (g/cc)	Comments
Polymethylpentene	PMP		0.835–0.840	Used in lab equipment and in cookware
Linear low-density polyethylene	LLDPE	4	0.915–0.950	Used in food packaging
Low density polyethylene	LDPE	4	0.917–0.940	
Ethylene vinyl acetate	EVA		0.920–0.940	Survives low temperatures well
High-density polyethylene	HDPE		0.940–0.970	
Polypropylene	PP	5	0.970–1.050	Large number of uses

It is noteworthy that the densities shown in Table 17.4 all appear to be rather close to each other. HDPE, for example, is close to low-density polyethylene (LDPE). There are plastics that have higher densities though, some of them significantly so. Polytetrafluoroethylene, for example, commonly called Teflon®, has a density of 2.10–2.20 g/cc.

Many plastics, including many we have discussed in previous chapters, find applications precisely because they are lightweight. Perhaps, the most obvious example is LDPE. Its uses include:

– Trash bags
– Packaging films
– Cabling insulation
– Deformable bottles
– Housewares

As mentioned previously though, other properties are also important in deciding upon a specific plastic and a specific application. For example, trash bags made from LDPE must be deformable – be able to stretch without breaking – as well as low density.

17.4 Recycling

Virtually all lightweight materials that are metals or metal alloys are recycled, as discussed in Chapter 8. Recycling programs exist for consumer end use items, such as beverage cans, but scrap yards recycle the metal components of automobiles and trucks, and specialized scrap yards exist for the disassembly, recycling, and reuse even of aircraft, and of large, ocean-going ships. These latter types of recycling are not seen often by the general public, yet still result in the reuse of a significant tonnage of iron, steel, copper, aluminum, and several other metals each year.

Plastics are also recycled extensively, as discussed in Chapter 7, but not based solely on their density. The LDPE mentioned in Table 17.3 is produced in large enough amounts that it has the resin identification code 4. The general public tends to think of plastic recycling in terms of recycling plastic beverage containers, but many plastic objects that were used in some other, possibly industrial, application are recycled as well.

The other side of plastics recycling is what is not recycled. The example we saw earlier in the chapter, Styrofoam®, is one of the plastics with a density so low that it is not profitable to transport it to recycling sites. Unfortunately, this means that almost all such materials, be they drinking cups or packaging inserts, are simply landfilled.

References

[1]   ELAC – European Lightweight Cluster Alliance. Website. (Accessed 16 June 2025, as: https://elcanet work.eu).

[2]   U.S. Office of Energy Efficiency & Renewable Energy. "Lightweight materials for cars and trucks." Website. (Accessed 16 June 2025 as: https://www.energy.gov/eere/vehicles/lightweight-materials-cars-and-trucks).

[3]   M. Mecklenburg, A. Schuchardt, Y. Kumar-Mishra, S. Kaps, R. Adelung, A. Lotnyk, L. Kienle, K. Schulte. "Aerographite: Ultra lightweight, flexible nanowall, carbon microtube material with outstanding mechanical performance", *Advanced Materials*, 2012, 24(26): 3486–3490.

[4]   Metal Foams: Fundamentals and Applications. N. Dukhan, 2012, ISBN: 978-16059-5014-3.

[5]   USGS Mineral Commodity Summaries, 2024. Downloadable.

[6]   Concrete Canoe. Website. (Accessed, 16 June 2025, as: https://www.concretecanoe.org).

Chapter 18
Hard materials

18.1 Introduction

Previous chapters in this book have discussed metals and alloys, glasses and ceramics, and cement and concrete. Many of these qualify as hard materials. As Chapter 17 examined materials in terms only of their density, this chapter looks at the broad classes of materials again, but this time with an emphasis specifically on their hardness, durability, and deformability. Numerous applications now exist for some material based simply on its hardness, which often equates to its ability to withstand stresses, strains, and impacts.

We have said in more than one of the previous chapters that a subject might be one considered to have fuzzy edges. This can be said of hard materials as well. There are certainly well-known hard materials, diamond being one example (that we discuss below). But there are several other materials that have been developed to a high degree rather quickly, often as a result of some development that occurred during the Second World War.

Two cases of novel hard materials that may seem apparent to most people are: first, the development of hard materials for engine components, and second, the continued development of vehicle and personal armor. One case that is probably not as well known is the development of new cutting abrasives. In the first case, engines in automobiles, ships, and planes have required new, hard materials because they have to function under conditions of much greater stress than before. In the second case, what is called the "lethality of warfare" has increased – meaning projectiles have enhanced penetrating power – and thus defensive measures must increase in effectiveness as well. In the third case, hard materials must still be shaped and cut, which in turn requires more effective cutting tools and materials.

As we examine a variety of hard materials, it is convenient to discuss how hardness is determined. Despite a great deal of research into hard materials and how they are made, there is no universally accepted means by which hardness is quantified. Many people are familiar with the Mohs hardness scale for minerals, which is still taught in many schools. Table 18.1 is a summary of the Mohs scale. In it, any harder mineral can scratch one with a lower number.

A mineral as "soft" as gypsum can be scratched with a fingernail. Indeed, the Mohs measure of hardness depends on the ability of a material to scratch or be scratched, not on how it can be deformed by direct pressure. Still, this softness makes some minerals easy to work and shape for both useful and decorative objects. Figure 18.1, as just one example, is a letter opener made of selenite, a form of gypsum.

Figure 18.2 illustrates a piece of calcite in such a way that the angles of the crystal can be seen. Calcite – calcium carbonate or $CaCO_3$ – is a relatively soft mineral.

https://doi.org/10.1515/9783112205822-018

No.	Example	Comments
1	Talc	Softest mineral, $Mg_3Si_4O_{10}(OH)_2$
2	Gypsum	$CaSO_4 \cdot 2H_2O$, example in Figure 18.1, can scratch talc
3	Calcite	$CaCO_3$, see in Figure 18.2
4	Fluorite	CaF_2
5	Apatite	A group of minerals, generally, $Ca_5(PO_4)_3(F,Cl,OH)$
6	Orthoclase	$KAlSi_3O_8$
7	Quartz	Base material for many ceramics and glasses
8	Topaz	Generally, $Al_2SiO_4(F,OH)_2$
9	Corundum	An aluminum oxide
10	Diamond	C_{10}, hardest naturally occurring material

Figure 18.1: Selenite letter opener.

There are several other hardness scales as well, including the:
- Brinell scale
- Knoop microhardness
- Rockwell hardness
- Vickers microhardness

Each has been developed experimentally and not as the end result of an overarching theory of the hardness of materials.

As an example of how the hardness of a material is determined, we will use the Vickers scale since it is straightforward and widely used. In it, a diamond pyramid of specific shape and angle, called the indenter, is pressed into the material. The indentation that is left in the material is then measured. A schematic of the Vickers indenter

Figure 18.2: Calcite crystal.

is shown in Figure 18.3. Other hardness test often involve pressing some point of known dimensions, such as a ball point, into the material to be tested.

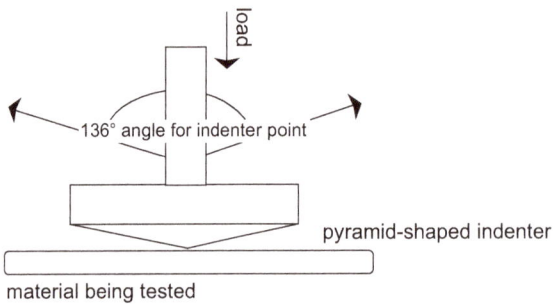

Figure 18.3: Vickers hardness indenter.

18.2 Basic classifications

A wide variety of materials find some use precisely because they are hard and usually also because they are durable. The following sub-sections discuss predominantly those which have some established application or applications; but other, less well-known ones are being developed as well.

18.2.1 Natural diamond

In numerous books and websites, it is stated that diamond is the hardest substance in the world. Less commonly stated is that it is brittle enough that it does not deform and thus can be made to shatter. This allotrope of carbon (C_{10}) is one of four known, the others being:
- Amorphous carbon
- Graphite (C_6)
- Fullerenes (the smallest being C_{60})

The smallest repeating unit structure for diamond is shown in Figure 18.4. It can be helpful to think of this structure as that of four carbon-based tetrahedra, each connected to four other carbon atoms. In turn, each of these is connected to four other carbon atoms in an almost infinite array.

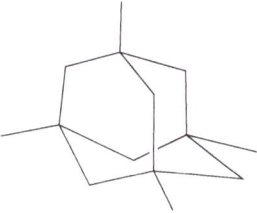

Figure 18.4: Diamond structure.

Macroscopically, diamond is the one allotrope of carbon that continues to evoke a certain wonder among people in many cultures. Traditionally, diamonds were found in alluvial deposits in several of the great rivers of the world. The mining or gathering of alluvial diamonds has not completely faded away as deep diamond mining and the production of synthetic diamonds have grown. But the majority of the natural diamonds used today are those that have been mined from deep deposits [1–3].

When discussing diamonds, diamond chemistry, and diamond production, it is worth noting that historically, since diamonds have been mined or have been taken from alluvial deposits, the largest diamonds are the most valuable simply because they are extremely rare. Table 18.2 is a list of a small number of the world-famous, large diamonds. A full listing would be hundreds of entries long. A statement like this may seem contradictory to that just mentioned, of the rarity of large diamonds, but bear in mind that millions of diamonds have been traded and sold in the past century alone. Trade organizations devoted to diamonds exist, advocating for their use, sale and ownership, while government agencies track such finds [1–4].

Small diamonds, often the size of a grain of sand, have been used in industry for decades – indeed, such diamonds are often called "industrials." They find use as abrasives in a wide variety of applications. They are a commodity tracked by the USGS Mineral Commodity Summaries as "Diamond (Industrial)," in which the comment is made:

Table 18.2: Large, natural diamonds.

Name	Date of discovery	Carat wt (uncut)	Comments
Cullinan	1905	3,106.75	Largest ever found
DeBeers	1888	428.5	
Hope		45.52 (cut)	According to legend, cursed
Jubilee	1895	650.8	
Koh-i-Noor		793	Now in the British crown jewels
Millenium Star	1990	777	Flawless
Star of Sierra Leone	1973	969	
Queen of Kalahari	2015	341.9	

"The major consuming sectors of industrial diamond are computer chip production; construction; drilling for minerals, natural gas, and oil; machinery manufacturing; stone cutting and polishing; and transportation (infrastructure and vehicles)" [5].

Clearly, the importance of industrial size and grade diamonds is high. Production of diamonds by various nations is tracked in millions of carats according to the Mineral Commodity Summaries.

Diamond is the only mineral commodity measured in carats (as opposed to kilograms or metric tons, for example), and it is sometimes helpful to convert this into a metric unit that is easier to recognize. For example, the 54 million carats reported in 2019 convert to 10,800 kg (approximately 23,810 pounds). This gives some idea of the scope of diamond production and use worldwide.

18.2.2 Diamond and silicon carbide

As mentioned, small, industrial diamonds – often the size of a grain of sand – have found uses as cutting materials simply because they are extremely hard and do not wear down, abrade, or wear away easily. Often called bort, these can be attached to a "sandpaper" surface or can be used as a free powder.

Silicon carbide (formula SiC), also called carborundum, has in the past been called the "poor man's diamond," simply because it is much more common than diamond. It can be found in nature as the mineral moissanite, but has been produced as an industrial material for more than a century [6, 7].

It is difficult to write stoichiometric reactions for the production of silicon carbide, but the following are the conditions routinely used:
1. Combine coke in powder form and sand (SiO_2) in an Acheson furnace (in an iron bowl).
2. Temperatures should be 1,600–2,500 °C.
3. As SiC forms, CO is given off.

Because the technique by which silicon carbide is made, the Acheson electric furnace requires a great deal of energy; in several cases for large-scale production, such an operation is located in close proximity to some source of electricity. The first of these plants to be constructed was near the Niagra Falls, in New York State, because of the existing hydroelectric power plant there. A schematic of an Acheson furnace is shown in Figure 18.5. The electrodes pass through the furnace wall into the raw mixture. With both heat and electrical current, the silicon carbide is formed.

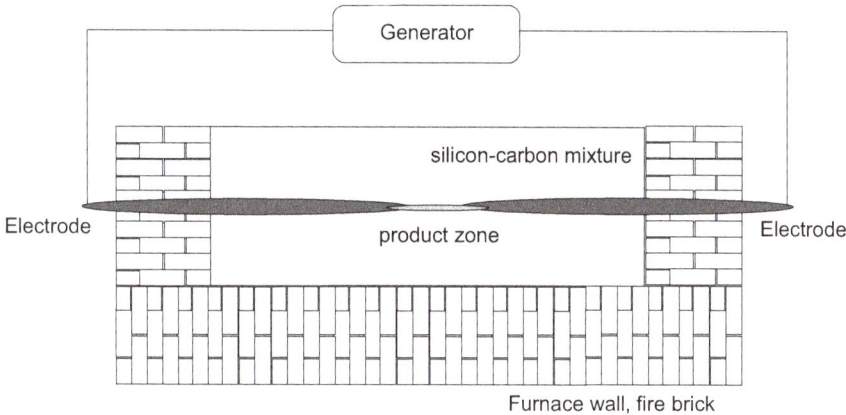

Figure 18.5: Acheson electrical furnace.

We have mentioned that silicon carbide is routinely used as a substitute for diamond – this is definitely so in the case of abrasives. The importance of this can be seen in that the United States Department of Defense tracks the use of it because it has strategic value [8]. It is one among the list of 76 materials all valued "because of their important defense uses and possible fragility of supply" [8].

Several different companies now produce silicon carbide. Besides abrasives, the material finds use as:

- anti-skid materials
- advanced ceramics
- brake pads
- buffing compounds
- glass
- refractory material coatings
- cosmetics

This last application may seem far removed from the others, but silicon carbide can be used as an abrasive in some dental applications.

18.2.3 Ceramic materials

Hard ceramics have found uses in the automotive industry, often as catalyst supports. The catalytic convertors in automobiles routinely depend on expensive platinum group metals, such as platinum or palladium. To maximize the surface area of such metals, it is common to deposit them on a ceramic support. The support can be a metal oxide, but must be hard enough to withstand the stresses of high-speed road travel. In specialized uses, such as that of certain military vehicles, such supports must be able to withstand the stresses of cross-country travel at relatively high speeds.

Also, a use that is important for a smaller automotive part is that of a spark plug insulator. These can be made of alumina, but formulas can also be proprietary.

18.2.4 Metal materials, carbides

A wide variety of metal carbides exist, many of them being hard materials. Routinely, such materials are made by combining a metal powder or metal oxide with powdered carbon at high temperature ($\approx$1,500 °C or higher) in a vacuum. Many of them find uses as abrasives, or in cutting tools when they are powders. As pressed materials, they tend to be solids that are highly resistant to wear and abrasion.

Using just one example, one tungsten carbide, known industrially as, "Tungsten carbide 857," has a composition of:

85.7% WC
9.5% Ni
1.8% Ta
1.5% Ti
1.0% Nb
0.3% Cr

Numerous other tungsten carbides exist though. How much of any one that is produced is a function of how much of that material is used, and what the continuing demand for it is.

18.2.5 Boron nitrides

Another class of binary compounds that are extremely hard are the boron nitrides. As with silicon carbide, many of the uses of boron nitride are for situations in which high stress or high temperatures must be endured.

There are different crystalline forms of boron nitride, but the most commonly used is hexagonal boron nitride or h-BN. The synthetic sequence to produce boron nitrides is shown in Figure 18.6.

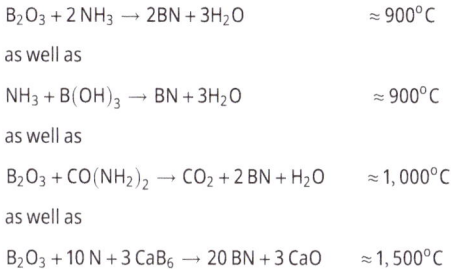

$$B_2O_3 + 2\,NH_3 \longrightarrow 2BN + 3H_2O \qquad\qquad \approx 900°C$$

as well as

$$NH_3 + B(OH)_3 \longrightarrow BN + 3H_2O \qquad\qquad \approx 900°C$$

as well as

$$B_2O_3 + CO(NH_2)_2 \longrightarrow CO_2 + 2\,BN + H_2O \qquad \approx 1,000°C$$

as well as

$$B_2O_3 + 10\,N + 3\,CaB_6 \longrightarrow 20\,BN + 3\,CaO \qquad \approx 1,500°C$$

Figure 18.6: Boron nitride synthesis.

Such reactions are run under elemental nitrogen atmosphere to minimize oxidation by-products, such as boron oxide, and to maximize purity, which can be as high as 99%. Should end-user items need to be made from such materials, as opposed to using them as powders, a high-temperature high-pressure system is utilized to force the material into the proper form.

Curiously, h-BN has found use as a lubricant, much like graphite. Further applications include the following:
- Abrasive dusts
- Alloy component
- Personal care products
- Paint component
- Ceramic additive
- Synthetic rubbers

A harder form of boron nitride is c-BN, which stands for cubic boron nitride. In structure, this form of boron nitride is close to diamond. One means of producing it is to subject h-BN to a high-temperature high-pressure chamber, approximately 10 GPa ($\approx$1.45 million psi) and 2,000–3,000 °C. It finds use in drills that are used in boring through steel because the material is entirely insoluble in iron, even at high temperatures.

18.2.6 Metal borides

Borides besides boron nitride exist as well: metal borides. In general, such materials are produced by the high-temperature combination of boron carbide, a metal oxide, and carbon, the reducing agent for the metal oxide [9]. The following are examples of those which have been produced:

Rhenium boride (ReB_2) – several possible stoichiometric formulas RuB_2;

OsB_2 – vacuum synthesis prevents formation of OsO_4;

ZrB_2 – considered a ceramic material. Often hot-pressed to produce some end item.

There are other metal borides besides these four. The field continues to be studied, and progress continues to be made [10–12] although how applicable some of these materials can become certainly remains in question. Osmium, for example, is such a rare element that one can question whether or not enough osmium boride can ever be made that it could become useful in some wide application. To make a comparison, osmium tetroxide has a long history of being a reagent that delivers two oxygen atoms to the cis-side of a double bond. But this reaction is no longer as widespread as it has been, or taught as much as it has been, simply because of the cost and scarcity of osmium.

Curiously, some of the metal borides, including ReB_2, have been found to exhibit electrical conductivity nearly on par with some of the metal elements. This may eventually lead to applications for the material.

18.3 Recycling

In general, the recycling of hard materials tends to be limited and to be concentrated on those materials which are used by the public. As mentioned in Chapter 15, glass recycling is an excellent example (since some types of glass can be considered a hard material); thus, glass bottles are recycled extensively in many areas. But whether hard or not, all glass transformations require high heat and thus can be recycled under roughly similar conditions. Catalytic converter recycling is also a well-established practice in many nations, so that the precious metal in the converter can be recovered and reused. The ceramic support used in such converters is not generally recycled into new converters, unless the process does not crush the existing object.

As well, synthetic diamonds, even materials containing small amounts of them, are recycled or reused simply because of the value of the dimonds. The driver for recycling is an economic one. The USGS Mineral Commodity Summaries does make comment about diamond recycling. It states:

In 2023, the amount of diamond bort, grit, and dust and powder recycled was estimated to be 14 million carats with an estimated value of $530,000. An estimated 75,000 carats of diamond stone was recycled with an estimated value of $110,000 [5].

To reiterate, the reason for this recycling is a straightforward one of economics.

References

[1]   World Diamond Council. Website. (Accessed 16 June 2025, as: https://www.worlddiamondcouncil.com/).
[2]   DeBeers Group. Website. (Accessed 16 June 2025, as: https://www.debeersgroup.com/).
[3]   The American Diamond Association. Website. (Accessed 16 June 2025, as: https://dmia.diamonds).
[4]   USGS Alluvial Diamond Resource Potential and Production Capacity Assessment of the Central African Republic. Downloadable as: https://pubs.usgs.gov/sir/2010/5043/pdf/sir2010-5043.pdf. (Accessed 16 June 2025).
[5]   United States Geological Survey, Mineral Commodity Summaries, 2024, downloadable.
[6]   Production of Artificial Carbonaceous Materials. E.G. Acheson US Patent 11,473, February 1895.
[7]   Electrical Furnace. E.G. Acheson. US Patent 560,291, May 1896.
[8]   Strategic and Critical Materials 2013 Report on Stockpile Requirements, Office of the Under Secretary of Defense for Acquisition, Technology and Logistics. Downloadable.
[9]   Method of Making Metal Borides. United States Patent: US 2957754A.
[10] Hard Materials – European Powder Metallurgy Association. Website. (Accessed 16 June 2025, as: https://www.epma.com).
[11] International Association of Advanced Materials. Website. (Accessed 16 June 2025, as: https://www.iaamonline.org).
[12] Materials Australia. Website. (Accessed 16 June 2025, as: https://www.materialsaustralia.com.au).

i.

Index

https://doi.org/10.1515/9783112205822-019

www.ingramcontent.com/pod-product-compliance
Lightning Source LLC
Chambersburg PA
CBHW061358210326
41598CB00035B/6021

* 9 7 8 3 1 1 9 1 4 7 3 9 2 *